Computational Approaches to Nuclear Receptors

RSC Drug Discovery Series

Editor-in-Chief
Professor David Thurston, *London School of Pharmacy, UK*

Series Editors:
Dr David Fox, *Pfizer Global Research and Development, Sandwich, UK*
Professor Salvatore Guccione, *University of Catania, Italy*
Professor Ana Martinez, *Instituto de Quimica Medica-CSIC, Spain*
Professor David Rotella, *Montclair State University, USA*

Advisor to the Board:
Professor Robin Ganellin, *University College London, UK*

Titles in the Series:
1: Metabolism, Pharmacokinetics and Toxicity of Functional Groups
2: Emerging Drugs and Targets for Alzheimer's Disease; Volume 1
3: Emerging Drugs and Targets for Alzheimer's Disease; Volume 2
4: Accounts in Drug Discovery
5: New Frontiers in Chemical Biology
6: Animal Models for Neurodegenerative Disease
7: Neurodegeneration
8: G Protein-Coupled Receptors
9: Pharmaceutical Process Development
10: Extracellular and Intracellular Signaling
11: New Synthetic Technologies in Medicinal Chemistry
12: New Horizons in Predictive Toxicology
13: Drug Design Strategies: Quantitative Approaches
14: Neglected Diseases and Drug Discovery
15: Biomedical Imaging
16: Pharmaceutical Salts and Cocrystals
17: Polyamine Drug Discovery
18: Proteinases as Drug Targets
19: Kinase Drug Discovery
20: Drug Design Strategies: Computational Techniques and Applications
21: Designing Multi-Target Drugs
22: Nanostructured Biomaterials for Overcoming Biological Barriers
23: Physico-Chemical and Computational Approaches to Drug Discovery
24: Biomarkers for Traumatic Brain Injury
25: Drug Discovery from Natural Products
26: Anti-Inflammatory Drug Discovery
27: New Therapeutic Strategies for Type 2 Diabetes: Small Molecules
28: Drug Discovery for Psychiatric Disorders
29: Organic Chemistry of Drug Degradation
30: Computational Approaches to Nuclear Receptors

How to obtain future titles on publication:
A standing order plan is available for this series. A standing order will bring delivery of each new volume immediately on publication.

For further information please contact:
Book Sales Department, Royal Society of Chemistry, Thomas Graham House, Science Park, Milton Road, Cambridge CB4 0WF, UK
Telephone: +44 (0)1223 420066, Fax: +44 (0)1223 420247
Email: booksales@rsc.org
Visit our website at http://www.rsc.org/Shop/Books/

Computational Approaches to Nuclear Receptors

Edited by

Pietro Cozzini
Molecular Modelling Laboratory, Department of General and Inorganic Chemistry, University of Parma, Parma, Italy
Email: pietro.cozzini@unipr.it

Glen E. Kellogg
Virginia Biotech 1, Virginia Commonwealth University, Richmond, VA, USA

RSCPublishing

RSC Drug Discovery Series No. 30

ISBN: 978-1-84973-364-9
ISSN: 2041-3203

A catalogue record for this book is available from the British Library

Published by The Royal Society of Chemistry,
Thomas Graham House, Science Park, Milton Road,
Cambridge CB4 0WF, UK

Registered Charity Number 207890

For further information see our web site at www.rsc.org
Printed in the United Kingdom by Henry Ling Limited, Dorchester DT1 1HD, UK

Preface

Nuclear receptors (NRs) are evolutionary conserved proteins that are expressed in the animal kingdom. The wide-ranging family of NRs contains more than 150 members that are represented by both steroidal and non-steroidal receptors and by a multitude of orphan receptors, whose physiologically relevant ligands have not yet been identified. Many of these proteins act, in fact, as intracellular ligand-inducible transcription factors able to respond to endogenous and exogenous chemicals regulating gene expression. These factors affect a surprisingly diverse variety of functions, such as reproductive development, detoxification of foreign substances and fatty acid metabolism.

While many proteins are defined as 'druggable targets', *i.e.*, proteins for which the biological function of a significant fraction of family members have been successfully modulated by small molecule drugs, NRs perhaps should be thought of as nutriable targets since they respond to intracellular metabolites and xenobiotics such as drugs and dietary agents. Both metabolites and xenobiotics bind to the E domain located at the carboxy terminus of the receptors, often referred to as the *ligand binding domain* (LBD), whose sequence varies substantially between NRs, but whose structure is well conserved.

Of particular and increasing importance, molecules able to bind NRs are defined as *endocrine-disrupting chemicals* (EDCs) since they are able to interfere with the physiological function of hormonal systems, thus affecting the health of humans and also domesticated and wild animals, causing developmental, reproductive and metabolic diseases. Although many EDCs are xenobiotics, derived by industrial processes and released into the environment (*i.e.*, pesticides, plasticizers, flame retardants, organotins, alkylphenols, dioxins, polychlorinated biphenyls, *etc.*) or present in food as

RSC Drug Discovery Series No. 30
Computational Approaches to Nuclear Receptors
Edited by Pietro Cozzini and Glen E. Kellogg
© The Royal Society of Chemistry 2012
Published by the Royal Society of Chemistry, www.rsc.org

flavoring, coloring, preservative or other additives, many others are naturally occurring (*e.g.*, phyto- and mycoestrogens) that are present in plants or fungi as part of their defensive mechanism against physical and biological stresses. Whether exposure to the EDCs is *via* the environment or *via* the food chain, it is usually persistent and could be potentially harmful, as in the case of xenobiotics, and/or beneficial, as for the phytoestrogens.

Industrialized and agricultural areas are typically polluted with a wide range of chemicals spread into air, soil and groundwater. Hence there are many people who are particularly and largely irreversibly exposed to these toxic compounds with higher risks for developing reproductive and/or endocrine dysfunctions. By dietary and environmental exposure, EDCs can affect the endocrine system by (i) mimicking natural hormones, (ii) antagonizing their action or (iii) modifying their synthesis, metabolism and transport. Most of the reported harmful effects of EDCs are attributed to their interference with NR-, hormone-like-mediated signaling.

However, the understanding of NRs on a molecular level is made much more difficult by their intrinsic flexibility and plasticity. The LBD binding sites are particularly plastic and dynamic, often dramatically changing conformation upon binding of their ligand. This affects both experimental approaches and computational approaches that probe (and exploit for drug discovery, *etc.*) the structure of nuclear receptors. Our goal in this book is to give to the reader a collection of viewpoints on the study of NRs with computational approaches. It is far beyond the scope of this volume to supply exhaustive answers to all the questions about NRs – these are extremely complex systems!

In the first chapter, Lorenzetti and Narciso depict the functions of NRs and the links between the environment and human organisms to show how NRs can act as mediators of the action of endocrine-active compounds.

In Chapter 2, Safo, Kellogg and Cozzini discuss how to use the available structural data as a platform for computational studies on NRs by reviewing the available data and explaining the benefits and limitations of these data with respect to the suite of computational approaches.

The following chapter, by Dal Palù, Dovier, Fogolari and Pontelli, describes how constraint logic programming can be used as an *ab initio* technique to study the structure and the flexibility of proteins. Here, it is applied to NRs.

Next, Spyrakis, Barril and Luque show the utility of molecular dynamics techniques in understanding the movements of NRs, in both high- and low-flexibility domains.

As NRs are very important targets for drugs and food additives, in Chapter 5, Abagyan, Chen and Kufareva describe screening/docking approaches for the prediction of small-molecule binding to NRs.

Bosma and Appell introduce quantum chemical approaches for studying estrogenic toxins in Chapter 6. They also describe how quantum mechanical approaches can be used to generate QSAR descriptors for virtual screening of drugs and food additives.

Arnatt and Zhang's chapter illustrates an example of homology modeling to build a structure model for further computational investigations. In particular, they attempt to characterize the ligand binding mode in the G protein-coupled estrogen receptor (GPER).

Because computational predictions can only be truly validated with real 'wet' experiments, the chapter by Della Torre and Maggi describes applications of *in vivo* tests using reporter bioluminescent mice to test and verify *in silico* predictions.

Finally, in the last chapter, Dall'Asta, Faccini and Galaverna illustrate an important application of using computational discovery tools. They show how with *in silico* predictions at molecular models for estrogen receptors, new potential pollutants can be identified and, with these predictions, new chemosensors for food safety can be developed.

Pietro Cozzini
Glen E. Kellogg

Contents

RSC Drug Discovery Series No. 30
Computational Approaches to Nuclear Receptors
Edited by Pietro Cozzini and Glen E. Kellogg
© The Royal Society of Chemistry 2012
Published by the Royal Society of Chemistry, www.rsc.org

CHAPTER 1

Nuclear Receptors: Connecting Human Health to the Environment

STEFANO LORENZETTI* AND LAURA NARCISO

Istituto Superiore di Sanità – ISS, Department of Food Safety and Veterinary Public Health, Food and Veterinary Toxicology Unit, viale Regina Elena 299, 00161 Rome, Italy
*E-mail: stefano.lorenzetti@iss.it

1.1 Introducing Nuclear Receptors

Nuclear receptors (NRs) are evolutionary conserved proteins whose encoding genes are expressed in the animal kingdom (metazoans); they are also present in animals that do not have any endocrine system.[1] NRs function as transcription factors activated by small (<1000 Da) lipophilic compounds able to cross the plasma membrane and, owing to their discovery as mediators of the sex steroid hormones, they were initially defined as endocrine receptors, although recently some of them have been suggested to act as sensors of their environment interacting with ligands external to the host organism (xenobiotics). Indeed, such ligand-activated transcription factors control levels of both xenobiotics (*i.e.*, man-made chemicals such as pesticides and plasticizers) and endobiotics (*i.e.*, endogenous chemicals such as sex steroid and thyroid hormones, vitamins) since the xenosensing activity of the NRs developed evolutionarily to support the spread of metazoans during the Cambrian age through the setting up of a whole endocrine system.[1] The survival of all organisms (*i.e.*, metazoans) relies on energy maintenance (*via* dietary intake, storage and utilization) and self-propagation (*via* reproduction), two physiological activities completely controlled by the central nervous system

RSC Drug Discovery Series No. 30
Computational Approaches to Nuclear Receptors
Edited by Pietro Cozzini and Glen E. Kellogg
© The Royal Society of Chemistry 2012
Published by the Royal Society of Chemistry, www.rsc.org

(CNS) through the signaling to the peripheral effector tissues/organs: NRs allow different, multiple signals to be integrated between central and peripheral organs acting as xenosensors and orchestrating hormone-dependent signaling.

1.2 Linking the Environment to the Human Organism: Nuclear Receptors as Mediators of the Action of Endocrine-active Compounds (EACs)

Ligands of NRs are usually defined as endocrine-active compounds (EACs) or endocrine-disrupting chemicals (EDCs), substances able to interfere with the function of hormonal systems affecting human and wild-life health, for example, contributing to developmental, reproductive and metabolic diseases.[2-5] Although many EACs are xenobiotics – man-made chemicals manufactured by industry and released into the environment (*e.g.*, pesticides, plasticizers, flame retardants, organotins, alkylphenols dioxins, polychlorinated biphenyls) – many others are naturally occurring (*e.g.*, phyto- and mycoestrogens), being present in plants or fungi as part of their defensive mechanism against physical and biological stresses. Either *via* the environment or *via* the food chain, exposure to EACs is usually persistent and common to a wide range of compounds whose role might be potentially both harmful (*e.g.*, xenobiotics) and/or beneficial (*e.g.*, phytoestrogens). Indeed, industrialized and agricultural areas are typically polluted with a wide range of chemicals spread into the air, soil and groundwater, hence people working with (or in some cases living near sources of) pesticides, fungicides and other man-made chemicals are particularly exposed to these toxic compounds and thus have a higher risk of developing reproductive and/or endocrine dysfunction. With dietary and environmental exposure, EACs can affect the endocrine system by (i) mimicking natural hormones, (ii) antagonizing their action or (iii) modifying their synthesis, metabolism and transport. Most of the reported harmful effects of EACs are attributed to their interference with hormone-like, NR-mediated signaling.[2,3]

1.3 How Nuclear Receptors Work

As mentioned above, human NRs (Table 1.1) constitute a superfamily of 48 ligand-activated transcription factors able to regulate cognate gene networks involved in key physiological functions such as cell growth and differentiation, development, homeostasis or metabolism.[3] NRs have a fairly simple and conserved general structure and, as depicted in Figure 1.1a, is constituted by five distinct domains characterized by subdomains having specific functions.[2] The modulatory A/B domain, at the amino terminus of each NR, contains the transcriptional activation function (AF-1) that, together with the AF-2 domain, takes part in receptor dimerization, nuclear localization and binding to co-activators and co-repressors (Figure 1.1b).[2,3,6] The C domain is highly conserved since overlaps with the DNA-binding domain (DBD), a region recognizing NR-specific response elements (NR-RE) in the promoter sequences of targeted genes.[2,3,6] The DBD contains two zinc fingers and TA boxes: the first zinc finger has a highly conserved sequence, termed P-box, involved in NR–DNA helix binding, whereas the second zinc finger contains a D-box necessary for protein–protein interactions and also for NR binding to the NR-RE. Finally, the TA boxes within the DBD are essential in NRs acting as monomers.[7] NR binding specificity relies on the orientation of the binding sites (the single consensus binding site being the sequence AGGTCA) that can be formed by AGGTCA inverted, reverted or direct repeats, and also on their separation (from one to eight nucleotides) between the two single consensus binding sites. Indeed, whereas some NRs bind with high stringency to the consensus binding site, others display greater flexibility.[2,3,6,7] The D domain is a short hinge region allowing each NR to undergo conformational changes upon ligand binding. The E–F domain, close to the carboxy terminus, contains the ligand-binding domain (LBD), a region containing the AF-2 domain that, as mentioned above, participates with the AF-1 domain in receptor dimerization, nuclear localization and binding to co-activators and co-repressors. The LBD is structured in α-helices that provide conformational flexibility to this region, allowing the DBD–LBD interaction and thus contributing also to the recruitment of co-regulators essential for the initiation of the transcription of target genes.[2,3,6,7]

In the absence of the cognate ligand, some NRs are located in the nucleus, bind to the DNA response elements of their target genes and recruit co-repressors, whereas others are located in the cytoplasm in an inactive complex with chaperones (Figure 1.1b).[2,3,6,7] Ligand binding induces major structural alterations of the receptor LBDs, leading to (i) destabilization of co-repressor or chaperone interfaces, (ii) exposure of nuclear localization signals to allow nuclear translocation and DNA binding of cytoplasmic receptors and (iii) recruitment of co-activators triggering gene transcription through chromatin remodeling and activation of the general transcription machinery.[3] The crystal structures of most LBDs of the NRs have been determined, revealing a conserved core of 12 α-helices (H1–H12) and a short two-stranded antiparallel β-sheet arranged into a three-layered sandwich fold; this arrangement generates a mostly hydrophobic cavity in the lower half of the domain which can accommodate the cognate ligand. In all ligand-bound LBD structures, the ligand binding pocket is sealed by the helix H12. This conformation is specifically induced by the binding of cognate ligands (endo- or xenobiotic) and is referred to as the 'active conformation' because it allows the dissociation of co-repressors and favors the recruitment of transcriptional co-activators.[2,3,6,7,8]

It is noteworthy that this conformational state can also be achieved by some constitutively active orphan receptors for which no natural ligands have been identified. In this active-form, helices H3, H4 and H12 define a hydrophobic binding groove for short LxxLL helical motifs (L represents leucine and x any

Table 1.1 List of the 48 human nuclear receptors (NRs) with their known ligands, common names and aliases, and also their official nomenclature. The existence of a crystallographic structure is indicated. A link to human diseases or pathophysiological conditions is also indicated for each NR.

No.	Name	Acronym (and aliases)	NR nomenclature	Crystallography PDB database (March 2010)	Dimers (m, d, RXRh)[a]	Ligands	Target diseases or pathophysiological conditions (PubMed search: Jan. 2012)
1	Thyroid hormone receptor α	THRA (AR7, EAR7, ERB-T-1, ERBA, ERBA1, THRA1, THRA2, c-ERBA-1)	NR1A1	Yes	RXR-h	Thyroid hormones	Hyperthyroidism, energy balance and metabolic homeostasis, cancer, cardiac failure
2	Thyroid hormone receptor β	THRB (ERBA-BETA, ERBA2, GRTH, PRTH, THR1, THRB1, THRB2)	NR1A2	Yes	RXR-h	Thyroid hormones	Hyperthyroidism, energy balance and metabolic homeostasis, cancer, cardiac failure
3	Retinoic acid receptor α	RARA (RAR)	NR1B1	Yes	RXR-h	Retinoic acid	Hematopoiesis, acne, psoriasis, neurological diseases, cancer
4	Retinoic acid receptor β	RARB (HAP, RRB2)	NR1B2	Yes	RXR-h	Retinoic acid	Hematopoiesis, tumorigenesis
5	Retinoic acid receptor γ	RARG (RARC)	NR1B3	Yes	RXR-h	Retinoic acid	Hematopoiesis, inflammation, acne, psoriasis, cancer
6	Peroxisome proliferator-activated receptor α	PPARA (MGC2237, MGC2452, PPAR, PPARalpha, hPPAR)	NR1C1	Yes	RXR-h	Fatty acids, leukotriene B4, fibrates	Hyperlipidemia, arteriosclerosis, diabetes, inflammatory diseases, biological rhythm disorders, skin disease

Table 1.1 (*Continued*)

No.	Name	Acronym (and aliases)	NR nomenclature	Crystallography PDB database (March 2010)	Dimers (m, d, RXRh,)[a]	Ligands	Target diseases or pathophysiological conditions (PubMed search: Jan. 2012)
7	Peroxisome proliferator-activated receptor β/δ	PPARD (FAAR, NUC1, NUCI, NUCII, PPARB)	NR1C2	Yes	RXR-h	–	Obesity, skin disease
8	Peroxisome proliferator-activated receptor γ	PPARG (CIMT1, GLM1, PPARG1, PPARG2, PPARgamma)	NR1C3	Yes	RXR-h	Prostaglandin J2, thiazolidinediones	Hyperlipidemia, arteriosclerosis, diabetes, inflammatory diseases, cancer Alzheimer's disease, skin disease
9	Rev-Erb α	NR1D1 (EAR1, THRA1, THRAL, ear-1, hRev)	NR1D1	Yes	m, d	Heme	Biological rhythm disorders (insomnia, depression, jet lag), mood disorders, arteriosclerosis, inflammation
10	Rev-Erb β	NR1D2 (BD73, EAR-1R, RVR)	NR1D2	Yes	m, d	Heme	Inflammation
11	RAR-related orphan receptor A	RORA (DKFZp686M2414, MGC119326, MGC119329, ROR1, ROR2, ROR3, RZR-ALPHA, RZRA)	NR1F1	Yes	m	Cholesterol, Cholesteryl sulfate	Arteriosclerosis, bone metabolism, neurological diseases

Table 1.1 (*Continued*)

No.	Name	Acronym (and aliases)	NR nomenclature	Crystallography PDB database (March 2010)	Dimers (m, d, RXRh)[a]	Ligands	Target diseases or pathophysiological conditions (PubMed search: Jan. 2012)
12	RAR-related orphan receptor B	RORB (ROR-BETA, RZR-BETA, RZRB, bA133M9.1)	NR1F2	Yes	m	Retinoic acid	Biological rhythm disorders, arteriosclerosis
13	RAR-related orphan receptor C	RORC (RORG, RZR-GAMMA, RZRG, TOR)	NR1F3	Yes	m	Orphan	Thymopoiesis, lymph node organogenesis
14	Liver X receptor α; oxysterols receptor LXR-α	NR1H3, LXR-a (LXRA, RLD-1)	NR1H3	Yes	RXR-h	Oxysterols, T0901317, GW3965	Atherosclerosis, diabetes, Alzheimer's disease, skin disorders, reproductive disorders, cancer
15	Liver X receptor β; oxysterols receptor LXR-β	NR1H2, LXR-b (LXRB, NER, NER-I, RIP15, UNR)	NR1H2	Yes	RXR-h	Oxysterols, T0901317, GW3965	Atherosclerosis, diabetes, Alzheimer's disease, skin disorders, reproductive disorders, cancer
16	Farnesoid X receptor; bile acid receptor	NR1H4, FXR (BAR, HRR-1, HRR1, MGC163445, RIP14)	NR1H4	Yes	RXR-h	Bile acids, fexaramine, lanosterol	Liver diseases, bile salt homeostasis, metabolic disease, cancer
17	Vitamin D (1,25-dihydroxyvitamin D₃) receptor	VDR	NR1I1	Yes	RXR-h	Vitamin D, 1,25-dihydroxyvitamin D_3	Osteoporosis, hypercalcemia, cutaneous inflammation

Table 1.1 *(Continued)*

No.	Name	Acronym (and aliases)	NR nomenclature	Crystallography PDB database (March 2010)	Dimers (m, d, RXRh)[a]	Ligands	Target diseases or pathophysiological conditions (PubMed search: Jan. 2012)
18	Pregnane X receptor; steroid and xenobiotic receptor	NR1I2, PXR (BXR, ONR1, PAR, PAR1, PAR2, PARq, PRR, SAR, SXR)	NR1I2	Yes	RXR-h	Xenobiotics, 16α-cyanopregnenolone	Drug metabolism, poisoning, bile salts homeostasis, liver diseases, inflammatory bowel disease, osteoporosis
19	Constitutive androstane receptor	NR1I3, CAR (CAR1, MB67)	NR1I3	Yes	RXR-h	Xenobiotics	Drug metabolism, poisoning, liver diseases
20	Hepatocyte nuclear factor 4, α	HNF4A (HNF4, HNF4a7, HNF4a8, HNF4a9, HNF4alpha, MODY1, NR2A21, TCF14)	NR2A1	Yes	d	Orphan	Diabetes, metabolic homeostasis
21	Hepatocyte nuclear factor 4, γ	HNF4G	NR2A2, NR2A3	Yes	d	Orphan	Metabolic homeostasis
22	Retinoid X receptor, α	RXRA	NR2B1	Yes	d	Retinoic acids	Metabolic and cardiovascular disorders, cancer
23	Retinoid X receptor, β	RXRB (DAUDI6, H-2RIIBP, MGC1831, RCoR-1)	NR2B2	Yes	d	Retinoic acids	Metabolic and cardiovascular disorders, cancer

Table 1.1 (*Continued*)

No.	Name	Acronym (and aliases)	NR nomenclature	Crystallography PDB database (March 2010)	Dimers (m, d, RXRh)[a]	Ligands	Target diseases or pathophysiological conditions (PubMed search: Jan. 2012)
24	Retinoid X receptor, γ	RXRG (RXRC)	NR2B3	Yes	d	Retinoic acids	Metabolic and cardiovascular disorders, cancer
25	Testicular receptor 2	NR2C1, TR2	NR2C1	?	d, RXR-h	Orphan	Retinal degenerations
26	Testicular receptor 4	NR2C2, TR4 (TAK1, TR2R1, hTAK1)	NR2C2	?	d, RXR-h		–
27	Tailless homolog	NR2E1, TLX (TLL, XTLL)	NR2E1	?	m, d	Orphan	Psychiatric diseases
28	Photoreceptor-specific nuclear receptor	PNR (ESCS, MGC49976, RNR, RP37, rd7)	NR2E3	?	m, d	Orphan	Retinal degenerations
29	Chicken ovalbumin upstream promoter transcription factor I	NR2F1, COUP-TF I (EAR-3, EAR3, ERBAL3, SVP44, TCFCOUP1, TFCOUP1)	NR2F1	Yes	d, RXR-h	Orphan	Oxidative stress, cancer, multiple sclerosis
30	Chicken ovalbumin upstream promoter transcription factor II	NR2F2, COUP-TF II (ARP1, COUPTFB, SVP40, TFCOUP2)	NR2F2	Yes	d, RXR-h	Orphan	Adrenal diseases, cancer

Table 1.1 (*Continued*)

No.	Name	Acronym (and aliases)	NR nomenclature	Crystallography PDB database (March 2010)	Dimers (m, d, RXRh)[a]	Ligands	Target diseases or pathophysiological conditions (PubMed search: Jan. 2012)
31	V-erbA-related protein 2	NR2F6, EAR-2 (EAR2, ERBAL2)	NR2F6	?	m	Orphan	
32	estrogen receptor 1; ER-α	ESR1 (RPI-130E4.1, ER, ESR, ESRA, Era)	NR3A1	Yes	d	17β-Estradiol, tamoxifen, raloxifene	Osteoporosis, cardiovascular diseases, cancer, melanosis
33	estrogen receptor 2; ER-β	ESR2 (ESR-BETA, ESRB, ESTRB, Erb)	NR3A2	Yes	d	Various synthetic compounds	Osteoporosis, cancer, endometriosis
34	Estrogen-related receptor α	ESRRA (ERR1, ERRa, ERRalpha, ESRL1)	NR3B1	Yes	m, d	Orphan	Metabolic diseases, osteoporosis, cancer
35	Estrogen-related receptor β	ESRRB (DFNB35, ERR2, ERRb, ESRL2)	NR3B2	Yes	m, d	DES, 4-OH-tamoxifen	Metabolic diseases
36	Estrogen-related receptor γ	ESRRG (ERR3, ERRgamma, FLJ16023, KIAA0832)	NR3B3	Yes	m, d	DES, 4-OH-tamoxifen	Metabolic diseases
37	Glucocorticoid receptor	GR (GCCR, GCR, GRL)	NR3C1	Yes	d	Cortisol, dexamethasone, RU486	Metabolic homeostasis, psychiatric disorders, inflammatory bowel disease

Table 1.1 (*Continued*)

No.	Name	Acronym (and aliases)	NR nomenclature	Crystallography PDB database (March 2010)	Dimers (m, d, RXRh)[a]	Ligands	Target diseases or pathophysiological conditions (PubMed search: Jan. 2012)
38	Mineralocorticoid receptor	MR (FLJ41052, MCR, MLR, NR3C2VIT)	NR3C2	Yes	d	Aldosterone, spirolactone	Blood pressure, electrolyte and fluid homeostasis
39	Progesterone receptor	PGR, PR	NR3C3	Yes	d	Progesterone, medroxyprogesterone acetate, RU486	Neurodegenerative disorders, stroke, cancer
40	Androgen receptor	AR (RP11-383C12.1, AIS, DHTR, HUMARA, HYSP1, KD, SBMA, SMAX1, TFM)	NR3C4	Yes	d	Testosterone, flutamide	Gonad development, prostate diseases, cancer, polycystic ovarian disease, spinal bulbar muscular atrophy
41	NGFIB	NR4A1, NGFIB (GFRP1, HMR, MGC9485, N10, NAK-1, NP10, NUR77, TR3)	NR4A1	Yes	m, d, RXR-h	Orphan	Metabolic homeostasis, inflammation, multiple sclerosis
42	NGFI-B/nur77 beta-type transcription factor homolog	NR4A2, NURR1 (HZF-3, NOT, RNR1, TINUR)	NR4A2	Yes	m, RXR-h	Orphan	Metabolic homeostasis, inflammation
43	Neuron-derived orphan receptor 1	NR4A3, NOR1 (CHN, CSMF, MINOR, TEC)	NR4A3	Yes	m	Orphan	Metabolic homeostasis, inflammation

Table 1.1 (Continued)

No.	Name	Acronym (and aliases)	NR nomenclature	Crystallography PDB database (March 2010)	Dimers (m, d, RXRh)[a]	Ligands	Target diseases or pathophysiological conditions (PubMed search: Jan. 2012)
44	Steroidogenic factor 1	NR5A1, SF1 (AD4BP, ELP, FTZ1, FTZF1, POF7, SF-1)	NR5A1	Yes	m	Orphan	Adrenal diseases, gonad development, endometriosis, pituitary function
45	Liver receptor homolog-1	NR5A2, LRH-1 (B1F, B1F2, CPF, FTF, FTZ-F1, LRH1)	NR5A2	Yes	–	Orphan	Bile salt homeostasis, diabetes, inflammatory bowel disease, cancer
46	Germ cell nuclear factor	NR6A1, GCNF (GCNF1, NR61, RTR)	NR6A1	?	d	Orphan	–
47	DSS–AHC critical region on the X chromosome protein 1	NR0B1, DAX1 (AHC, AHCH, AHX, DAX-1, DSS, GTD, HHG)	NR0B1	Yes	–	Orphan	Adrenal diseases
48	Small heterodimer partner	NR0B2, SHP (FLJ17090, SHP, SHP1)	NR0B2	Yes	RXR-h	Orphan	Metabolic diseases

[a] m = monomer; d = dimer; h = heterodimer with RXR. Adapted from Germain *et al.* (2006)[81] and le Maire *et al.* (2010).[3]

(a)

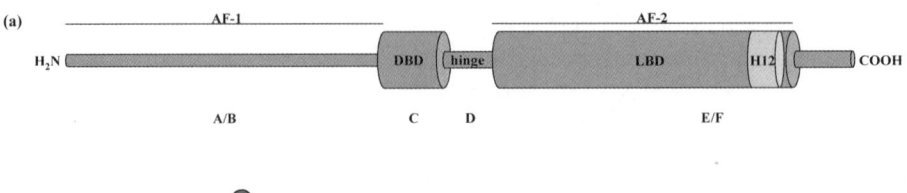

(b)

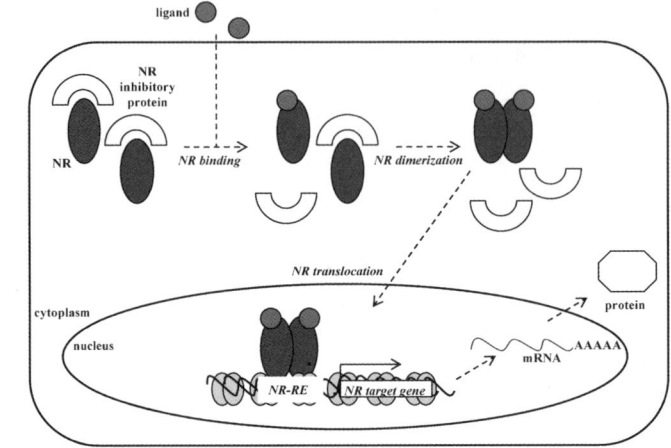

Figure 1.1   (a) Generic structure of a nuclear neceptor (NR) and (b) schematic
illustration of a typical NR-mediated signaling. (a) Overall structure of
NRs contain a well-conserved DNA-binding domain (DBD) and a
moderately conserved ligand-binding domain (LBD) separated by a
hinge region. In contrast, the N-terminal A/B region is highly divergent.
Two transcriptional activation functions have been described: a
'constitutively active' AF-1 in region A/B and an AF-2 which
corresponds to a coactivator binding surface formed by helices H3,
H4, and H12 of the LBD whose completion requires the presence of the
ligand. (b) NR-mediated signaling is usually activated by the ligand–NR
interaction within the cytoplasm that, upon displacing of the NR
inhibitory protein and its dimerization, leads to NR translocation within
the nucleus followed by (in)direct DNA binding and gene transcription
activation/inhibition. See the text and cited literature for further details.

amino acid) found within co-activators. In contrast to agonist binding,
interaction with antagonists prevents the correct positioning of helix H12, thus
avoiding association with the LxxLL motifs of co-activators.[8]

Since NRs form a complex set of interacting proteins that allow the body to
coordinate responses to fluctuations in chemical levels, it is obvious that they
undergo 'cross-talk'.[2] It is becoming increasingly clear that NRs interact
together and one of the great challenges is to ascertain how this interaction
network fully functions and to be able to predict what the biological response
will be for any given stimuli. NR 'cross-talking', indeed, has the twin
advantages both of ensuring the most efficient response to a given stimuli and
of providing a safety net to guarantee always an active capture system for a
stimulus, even in absence of the cognate receptor. Interactions between NRs

may occur at the level of sharing ligands and/or co-regulators and/or heterodimer partners and/or DNA binding elements. The best studied of these interactions is probably at the level of the target gene sets activated by NRs: the constitutive androstane receptor (CAR) and the pregnane-X-receptor (PXR), for example, coordinately regulate a battery of genes involved in all aspects of drug metabolism, including oxidative metabolism, conjugation and transport.[9,10] PXR and CAR, indeed, have been shown to regulate commonly several genes encoding drug-metabolizing enzymes and drug transporters, a common feature of their biology, but also genes involved in apparently completely different processes such as the karyopherin-mediated nuclear import.[11–13] The meaning of such coordinated interplay and the biological impact on the response to chemical stimuli are a challenge for future investigations on NR 'cross-talking'.

1.4 Environmental Chemicals and Adverse Effects on Human Health

In our environment, in consumer products and in foods, are commonly present chemicals, the above-mentioned EACs or EDCs, that interfere with hormone biosynthesis, metabolism or action resulting in a deviation from normal homeostatic control even during the developmental and reproductive life stages.[4] Indeed, they have been defined as EDCs because they often possess phenolic moieties that structurally mimic natural steroid hormones and enable EDCs to interact with steroid hormone receptors as (anti)antagonists. Most EDC actions and their role in pathophysiological conditions are exerted through NRs (see Table 1.1), including estrogen receptors (ERs), androgen receptor (AR) and thyroid receptors (TRs).

The link between EDC exposure and human health has been accepted worldwide following the suggestion of Sharpe and Skakkebaek that the increasing incidence of reproductive abnormalities in the human male may be related to increased *in utero* estrogen exposure.[14] This first speculation led subsequently to the definition of the so-called testicular dysgenesis syndrome (TDS), in which reproductive disorders of newborn (cryptorchidism, hypospadias) and young adult males (low sperm counts, testicular germ cell cancer), could be due to *in utero* exposure to EACs.[15] The incidence of these symptoms has increased in recent decades and may result from an irreversible developmental disorder originating in early fetal life.[14–17] This hypothesis has been supported by findings in animal models of TDS involving fetal exposure to phthalates.[16,17]

Overall, to understand the mechanism of action and consequences of exposure to EDCs, some important and general issues have to be taken into consideration, such as the age and latency of exposure, the low dose–response and mixture effects in addition to heritability.

Concerning both the age and latency of exposure, it is always important to keep in mind that the exposure of an adult to an EDC may have different consequences with respect to a developing fetus or infant. As an example, a recent review by Sharpe is noteworthy, in which the impact of environmental and lifestyle factors on spermatogenesis was reviewed considering the exposure to EACs/EDCs that may occur in fetal life and how this might then impact on spermatogenesis in adulthood.[18] Sharpe highlighted how exposure to environmental contaminants in perinatal life causes adverse effects on testis development that are irreversible (probably because fetal germ cells are affected), whereas the effects on the process of spermatogenesis in adulthood are probably reversible.[18] Hence the consequences of developmental exposure may not be immediately apparent early in life but may be manifested in adulthood or during aging.

Concerning the mixture issue, it is important to remember that, although the effects of contaminants are typically studied in single exposures, environmental exposures are rarely from a sole contaminant since organisms are exposed to different chemicals due to occupational, environmental, dietary or pharmaceutical exposure. As a result, mixtures of chemicals may have activities that are greater than that of one chemical alone (additive or synergistic effects) or less than predicted (antagonistic).[19,20] As an example, the work of Rider *et al.* on the cumulative effects of antiandrogenic chemicals on the reproductive development of the male rat upon *in utero* exposure to binary, ternary or more complex (up to seven) mixtures could be considered pivotal.[21] All tested combinations, using compounds that act by disparate mechanisms of toxicity, produced cumulative, dose-additive effects on the androgen-dependent tissues and, importantly, the chemicals were tested at concentrations below their individual "No Observable Adverse Effect Level" (NOAEL).[21,22]

Concerning the dose–response curve, it should be remembered that monotonic dose–response relationships are not always applicable to toxicants and in particular to EACs/EDCs, which are recognized to give, for instance, inverted-U dose–response relationships where very low doses stimulate growth and very high doses completely inhibit development.[23] Indeed, non-monotonic (or biphasic) dose–response relationships commonly occur in endocrinology for virtually all hormones and for EACs, as witnessed by the conclusion of a US Environmental Protection Agency expert panel that estrogenic chemicals can cause biological effects in laboratory animals at levels below those normally found to be safe, an assessment that runs counter to the conventional wisdom in toxicology.[24] Unexpected unique low-dose effects have been observed, for instance, for bisphenol A (BPA) stimulation of human prostate cell growth and also for the greater effect of pesticides on mice fertility at low but not high doses of the chemicals.[25,26]

Concerning heritability, one has to consider the so-called transgenerational phenotype, in which the EACs/EDCs can affect not only the exposed individual but also the children and subsequent generations: in this case, the adverse effect can be transmitted epigenetically (*i.e.*, *via* altered DNA methylation or histone acetylation or miRNA expression) through the germline even if the subsequent generations are not directly exposed to the environmental factors.[27] A well-known example of germline transmission is a rat model with the fungicide vinclozolin: embryonic exposure to vinclozolin during gonadal sex determination induced a transgenerational effect (from generation F1 to F4) on adult male reproduction and sperm production and also on breast tumors, prostate disease, kidney disease, immune abnormalities and metabolic disorders.[28,29]

1.5 Reproductive Diseases and Fertility

All over the world, human fertility rates are declining. In many Western countries, the fertility rate is below the replacement level and often this is attributed to socio-economic factors and increasing control of fertility, although the decreasing ability to conceive is appearing more and more influenced by environmental and lifestyle factors. The most common reproductive pathologies associated with the action of EDCs are clinically dimorphic: (i) male sexual differentiation is androgen dependent and the main associated diseases are the testicular dysgenesis syndrome (TDS) and prostate pathologies in which altered NR expression has been shown; (ii) female differentiation, in contrast, occurs largely independently of sex steroids and the main associated disorders are premature thelarche and pubarche, disorders of ovulation and lactation, breast diseases, endometriosis and uterine fibroids.[30–40]

1.5.1 Testicular Dysgenesis Syndrome and Spermatogenesis

Male infertility is closely linked to impaired gonadal development and function. Patients with testicular dysgenesis syndrome (TDS) present symptoms such as abnormal spermatogenesis, cryptorchidism (undescended testes), penile malformations (*e.g.*, hypospadias, a congenital malformation with abnormal placement of the external urethral orifice) and testicular cancer.[15–17] In patients with TDS-associated symptoms, a pathological role of Leydig cells has also been identified. The patients have higher levels of luteinizing hormone (LH) and follicle-stimulating hormone (FSH) and a tendency towards lower levels of testosterone. A decreased level of testosterone is correlated with age, but also other factors such as health, lifestyle and environmental factors.[41] Epidemiological studies with animal models exposed to phthalates showed how exposure to xenobiotics could lead to the male reproductive disorders associated with TDS.[16–17] The mechanism involved includes increased exposure to estrogens, decreased androgen production by suppression of Leydig cell function, suppression of the expression of the AR or altered androgen–estrogen balance.[42] Studies on the association between exposure to a specific chemical or a chemical class and risk of TDS showed little consistency. In contrast, several investigations showed an increased risk of TDS following maternal exposure to a range of persistent EDCs used in cosmetics/toiletries/medications during pregnancy.[15–17] Hence a possible explanation of these effects on the developing fetal testis is that they occurred as the result of exposure to a mixture of EDCs.

Overall, human fertility is remarkably low compared with most animals. Important features of low spermatogenesis are a low sperm count and a low semen quality in young men.[43,44] These observations suggest that there are fundamental differences between spermatogenesis in humans and other species. In addition, there is good evidence for considerable geographical variations in sperm count that indicate variations in environmental exposures and/or in genetic/ethnic influences.[43-45] The number of sperm produced in adulthood is critically determined by the number of Sertoli cells within the testis and proliferation of Sertoli cells is induced by testosterone produced by the Leydig cells: experimental rat studies showed that AR knockout results in about a 50% reduction in Sertoli cell numbers at birth.[46] Lifestyle and/or environmental factors acting during pregnancy could interfere with Sertoli and Leydig cells and induce effects on sperm counts in adulthood. Another factor involved in low sperm counts is smoking by mothers during pregnancy. A possible explanation involves the interaction of polycyclic aromatic hydro-carbons (PAHs) or other constituents in the cigarette smoke with the aryl hydrocarbon receptor (AhR), although other mechanisms are possible.[47]

Male infertility derived from occupational exposure is well documented. A pivotal study was conducted in California among male pesticide manufacturing workers exposed to 1,2-dibromo-3-chloropropane (DBCP), a nematocide used on banana and pineapple crops, possessing an estrogen-like activity that resulted in a selective effect on the seminiferous tubules.[48,49] Other phytochemicals and also plasticizers – acting mainly, but not only, *via* sex steroid receptors – have been shown to affect male fertility either at the level of spermatogenesis or at the level of the prostate gland, and include many pesticides and dioxins.[50-54]

In addition to phytochemicals, trace elements, often referred to as heavy metals, have also been shown to have an averse role in male reproduction, affecting semen quality and reproductive hormone levels.[55] Experimental and occupational studies pointed out a role for low-level exposure to cadmium, lead and mercury: all of them are suggested to alter the redox capability of spermatozoa, although at least cadmium is suspected to act either as a metallo-hormone or as an EAC itself.[56,57]

1.5.2   Female Dysfunctions and EACs/EDCs

Different studies support a role of EACs/EDCs in the pathogenesis of several female reproductive disorders, including polycystic ovarian syndrome (PCOS), aneuploidy, premature ovarian failure (POF), reproductive tract anomalies, uterine fibroids, endometriosis and ectopic gestation. Exposure to EACs/EDCs can contribute to the development of female reproductive disorders, particularly those occurring during a critical window of susceptibility, namely *in utero*, neonatally, in childhood, the puberty and adulthood.[4]

PCOS is a heterogeneous pathology characterized by persistent anovulation, oligo- or amenorrhea and hyper-androgenism in the absence of thyroid, pituitary, and/or adrenal diseases. Many patients have high levels of LH compared with FSH. This complex disorder likely has its origins both within and outside the hypothalamic–pituitary–ovarian axis; endocrine, neuroendocrine and metabolic regulators contribute to its manifestation. Studies in sheep and rhesus monkeys suggested that high levels of testosterone exposure induce symptoms of PCOS in adulthood such as anovulatory infertility, central adiposity, insulin resistance and hypersecretion of LH.[58] Some evidence suggests that *in utero* exposure of human female fetuses to androgen-like EDCs, and hence an altered AR-mediated signaling, could result in PCOS in adulthood.[59] Noteworthily, the involvement of the AR in PCOS seems to be mainly mediated by a its polyglutamine CAG (from 11 to 38) repeats: shorter alleles of the (CAG)*n* in exon 1 of the AR gene enhanced the susceptibility to PCOS, either by upregulating AR activity or by causing hyperandrogenism.[60] Interestingly, women affected by PCOS have higher levels of BPA and also of androgens: there is a strong relationship between serum BPA and androgen concentrations, speculatively due to the effect of androgen on the metabolism of BPA.[61]

Other pathways may be involved in endocrine disruption of PCOS: among numerous candidate genes that are associated with predisposition to developing PCOS in women there is also LH, LH receptor and PPARγ.[62] Indeed, the latter seems to be the real new player in PCOS and overall in placental and fetal development and in disturbed pregnancy. In addition to sex steroid receptors, PPARs have recently been focused upon as the mediators of PCOS in the association of this disorder with abdominal adiposity and insulin resistance, with the explanation that PCOS complex inheritance combines reproductive with metabolic abnormalities in which adipose tissue and insulin resistance may play an important role.[63]

POF is the cessation of proper ovarian function before the age of 40 years. Susceptibility during organogenesis and adult exposures to EACs/EDCs are possible causes of this pathology: EACs are likely to affect follicular stock in humans and/or animals, by direct toxicity towards follicles or by increasing their recruitment at each cycle.[64] Exposure *in utero* to diethylstilbestrol (DES), a well-known estrogen-like chemical acting *via* ERs, induced an earlier age of menopause, probably due – as evidenced in animal studies – to a decreased number of oocytes.[65,66] Noteworthily, animal studies suggested the pivotal importance of the AR in oocyte development and function: AR-null (ARKO) female mice are subfertile, possess a defective folliculogenesis and develop POF.[67–69] The latter have been shown to be due the lack of the AR within the granulose cells (GCs) and not within the oocytes, suggesting that AR signaling within the GCs of the ovary is a critical regulator of androgen-mediated follicle growth and development and, furthermore, that local androgen actions in ovarian GCs, rather than in the pituitary or hypothalamus, may be the primary site of androgen-regulated fertility in female mice.[70]

1.6   Obesity and Obesogens

During recent decades, a worldwide obesity increase among children, adolescents and adults has been assessed and recognized as one of the major global health problems since its health hazard is linked to a number of common diseases, including breast and prostate cancers. Increased adipose mass elevates the risk for the initiation or progression of a variety of pathological conditions related overall to the so-called metabolic syndrome: components of the metabolic syndrome – such as insulin resistance, hyperinsulinemia, hypertension and dyslipidemia – constitute major risk factors for the development of diabetes mellitus type 2 and coronary heart disease. Obesity occurs when energy intake exceeds energy expenditure but there is a huge variability in the propensity of each individual to gain weight and accrue fat mass, even at identical levels of excess caloric input. The 'environmental obesogen' hypothesis proposes that a subset of EACs/EDCs disrupts normal development or interferes with the body's homeostatic controls.[71,72] Thus, certain EACs/EDCs behave as obesogenic chemicals or obesogens and may interfere with the highly efficient homeostatic mechanisms regulating adipogenesis and energy balance.[71–74] Most obesogens can mimic lipophilic hormones and dietary fatty acids acting on the same metabolic sensors: NRs such as PPARs (peroxisome proliferator activated receptors), TRs (thyroid receptors), ERs and ERRs, RXRs (retinoid X receptors) and GR (glucorticoid receptor) are targets for transcriptional regulators that control metabolic homeostasis and as such may be the mediators of altered signaling leading to metabolic disorders.[71–74]

So far, the EACs/EDCs more frequently reported to act as obesogens are within the chemical classes of organotins [*e.g.*, tributyltin (TBT) and triphenyltin (TPT)], plasticizers (*e.g.*, BPA and phthalates) and surfactants [*e.g.*, perfluoroalkyl compounds (PFCs)].

TBT, for example, is a selective ligand of both RXRs and PPARs, whose heterodimers control adipocyte number, size and function. Prenatal exposure to TBT results in precocious lipid accumulation in adipose tissues and hepatosteatosis of newborn mice.[75] Long-term effects of prenatal exposure include an increase in epididymal fat mass and a trend towards body weight gain with age.[75] Also, phthalates and PFCs are agonists for one or more of the PPARs and are speculated to induce obesity or metabolic disorders acting through a non-PPARα-mediated pathway. Intrauterine exposure to di(2-ethylhexyl) phthalate (DEHP) has been shown to induce a delayed hepatocyte maturation and hepatosteatosis in F1 mice that are mainly mediated by activation of the pregnane X receptor (PXR).[76,77]

Interestingly, flavonoids such as genistein and daidzein (usually identified in dietary soy) acting *via* ER signaling reversed the fat accumulation in postmenopausal women.[78] Low concentrations of genistein acted as an estrogens and exerted an inhibitory effects on lipogenesis involving ERβ receptor, whereas high concentrations of genistein promoted lipogenesis through PPARγ, in an ER-independent mechanism.[79,80]

1.7 Conclusion

NRs control both the communication among the different organism tissues and the interaction between organisms and their surrounding environment. They act as xenosensors and endocrine regulators, thus playing a central role in connecting and integrating endogenous hormone-regulated functions to external dietary and/or environmental stimuli. Since all body tissues express a subset of NRs, overall NRs have pivotal control on the whole organism homeostasis and particularly on the pathophysiological status of an organism. Hence NRs are feasible therapeutic targets for dozens of human diseases and the discovery and development of compounds that finely modulate the activity of NRs may result in potential drugs only if a deeper knowledge of each ligand–NR interaction can be achieved.

References

1. F. M. Sladek, *Mol. Cell Endocrinol.*, 2011, **334**(1–2), 3–13.
2. N. Plant and S. Aouabdi, *Xenobiotica*, 2009, **39**(8), 597–605.
3. A. le Maire, W. Bourguet and P. Balaguer, *Cell. Mol. Life Sci.*, 2010, **67**(8), 1219–1237.
4. E. Diamanti-Kandarakis, J. P. Bourguignon, L. C. Giudice, R. Hauser, G. S. Prins, A. M. Soto, R. T. Zoeller and A. C. Gore, *Endocr. Rev.*, 2009, **30**(4), 293–342.
5. A. K. Hotchkiss, C. V. Rider, C. R. Blystone, V. S. Wilson, P. C. Hartig, G. T. Ankley, P. M. Foster, C. L. Gray and L. E. Gray, *Toxicol. Sci.*, 2008, **105**(2), 235–259.
6. H. Greschik and D. Moras, *Curr. Top. Med. Chem.*, 2003, **3**(14), 1573–1599.
7. A. Aranda and A. Pascual, *Physiol. Rev.*, 2001, **81**(3), 1269–1304.
8. J. P. Renaud and D. Moras, *Cell. Mol. Life Sci.*, 2000, **57**(12), 1748–1769.
9. S. Kakizaki, D. Takizawa, H. Tojima, N. Horiguchi, Y. Yamazaki and M. Mori, *Front. Biosci.*, 2011, **17**, 2988–3005.
10. C. Köhle and K. W. Bock, *Biochem. Pharmacol.*, 2009, **77**(4), 689–699.
11. A. Ueda, H. K. Hamadeh, H. K. Webb, Y. Yamamoto, T. Sueyoshi, C. A. Afshari, J. M. Lehmann and M. Negishi, *Mol. Pharmacol.*, 2002, **61**(1), 1–6.
12. J. M. Maglich, C. M. Stoltz, B. Goodwin, D. Hawkins-Brown, J. T. Moore and S. A. Kliewer, *Mol. Pharmacol.*, 2002, **62**(3), 638–646.
13. K. E. Plant, D. M. Everett, G. Gordon Gibson, J. Lyon and N. J. Plant, *Pharmacogenet. Genom.*, 2006, **16**(9), 647–658.
14. R. M. Sharpe and N. E. Skakkebaek, *Lancet*, 1993, **341**, 1392–1395.
15. R. M. Sharpe and N. E. Skakkebaek, *Fertil. Steril.*, 2008, **89**(2 Suppl.), e33–e38.
16. J. S. Fisher, *Reproduction*, 2004, **127**(3), 305–315.
17. S. H. Swan, K. M. Main, F. Liu, S. L. Stewart, R. L. Kruse, A. M. Calafat, C. S. Mao, J. B. Redmon, C. L. Ternand, S. Sullivan and J. L. Teague, *Environ. Health Perspect.*, 2005, **113**(8), 1056–1061.

18. R. M. Sharpe, *Philos. Trans. R. Soc. Lond. B Biol. Sci.*, 2010, **365**(1546), 1697–1712.
19. W. S. Baldwin and J. A. Roling, *Toxicol. Sci.*, 2009, **107**(1), 93–105.
20. I. Silins and J. Högberg, *Int. J. Environ. Res. Public Health*, 2011, **8**(3), 629–647.
21. C. V. Rider, V. S. Wilson, K. L. Howdeshell, A. K. Hotchkiss, J. R. Furr, C. R. Lambright and L. E. Gray Jr, *Toxicol. Pathol.*, 2009, **37**(1), 100–113.
22. C. V. Rider, J. R. Furr, V. S. Wilson and L. E. Gray Jr, *Int. J. Androl.*, 2010, **33**(2), 443–562.
23. W. V Welshons, S. C. Nagel and F. S. vom Saal, *Endocrinology*, 2006, **147**(6 Suppl.), S56–S69.
24. W. V. Welshons, K. A. Thayer, B. M. Judy, J. A. Taylor, E. M. Curran and F. S. vom Saal, *Environ. Health Perspect.*, 2003, **111**(8), 994–1006.
25. Y. B. Wetherill, C. E. Petre, K. R. Monk, A. Puga and K. E. Knudsen, *Mol. Cancer Ther.*, 2002, **1**(7), 515–524.
26. M. F. Cavieres, J. Jaeger and W. Porter, *Environ. Health Perspect.*, 2002, **110**(11), 1081–1085.
27. J. P. Bourguignon and A. S. Parent, *Curr. Opin. Pediatr.*, 2010, **22**(4), 470–477.
28. M. D. Anway and M. K. Skinner, *Endocrinology*, 2006, **147**(6 Suppl.), S43–S49.
29. M. D. Anway and M. K. Skinner, *Prostate*, 2008, **68**(5), 517–529.
30. U. N. Joensen, N. Jørgensen, E. Rajpert-De Meyts and N. E. Skakkebaek, *Basic Clin. Pharmacol. Toxicol.*, 2008, **102**(2), 155–161.
31. M. V. Maffini, B. S. Rubin, C. Sonnenschein and A. M. Soto, *Mol. Cell Endocrinol.*, 2006, **254–255**, 179–186.
32. I. Ceccarelli, D. Della Seta, P. Fiorenzani, F. Farabollini and A. M. Aloisi, *Neurotoxicol. Teratol.*, 2007, **29**(1), 108–115.
33. N. Atanassova, C. McKinnell, K. Williams, K. J. Turner, J. S. Fisher, P. T. Saunders, M. R. Millar and R. M. Sharpe, *Endocrinology*, 2001, **142**(2), 874–886.
34. C. Gupta, *Proc. Soc. Exp. Biol. Med.*, 2000, **224**(2), 61–68.
35. L. Ibanez and F. de Zegher, *Mol. Cell Endocrinol.*, 2006, **254–255**, 22–25.
36. I. Colon, D. Caro, C. J. Bourdony and O. Rosario, *Environ. Health Perspect.*, 2000, **108**(9), 895–900.
37. J. A. McLachlan, E. Simpson and M. Martin, *Best Pract. Res. Clin. Endocrinol. Metab.*, 2006, **20**(1), 63–75.
38. P. D. Darbre, *Best Pract. Res. Clin. Endocrinol. Metab.*, 2006, **20**(1), 121–143.
39. S. E. Fenton, *Endocrinology*, 2006, **147**(6 Suppl.), S18–S24.
40. R. R. Newbold, W. N. Jefferson and E. Padilla-Banks, *Reprod. Toxicol.*, 2007, **24**(2), 253–258.
41. S. Li, S. D. Hursting, B. J. Davis, J. A. McLachlan and J. C. Barrett, *Ann N. Y. Acad. Sci.*, 2003, **983**, 161–169.
42. R. M. Sharpe, *Int. J. Androl.*, 2003, **26**(1), 2–15.
43. N. Jørgensen, C. Asklund, E. Carlsen and N. E. Skakkebaek, *Int. J. Androl.*, 2006, **29**(1), 54–61.

44. A. M. Andersson, N. Jørgensen, K. M. Main, J. Toppari, E. Rajpert-DeMeyts, H. Leffers, A. Juul, T. K. Jensen and N. E. Skakkebaek, *Int. J. Androl.*, 2008, **31**(2), 74–80.
45. R. M. Sharpe, *Nat. Rev. Endocrinol.*, 2011, **7**(11), 633–634.
46. H. M. Scott, G. R. Hutchison, I. K. Mahood, N. Hallmark, M. Welsh, K. De Gendt, G. Verhoeven, P. O'Shaughnessy and R. M. Sharpe, *Endocrinology*, 2007, **148**(5), 2027–2036.
47. C. H. Ramlau-Hansen, A. M. Thulstrup, L. Storgaard, G. Toft, J. Olsen and J. P. Bonde, *Am. J. Epidemiol.*, 2007, **165**(12), 1372–1379.
48. D. Whorton, R. M. Krauss, S. Marshall and T. H. Milby, *Lancet*, 1977, **ii**(8051), 1259–1261.
49. Y. J. Oh, Y. J. Jung, J. W. Kang and Y. S. Yoo, *Sci. Total Environ.*, 2007, **388**(1–3), 8–15.
50. R. Hauser and R. Sokol, *Fertil. Steril.*, 2008, **89**(2 Suppl.), e59–e65.
51. F. Orton, E. Rosivatz, M. Scholze and A. Kortenkamp, *Environ. Health Perspect.*, 2011, **119**(6), 794–800.
52. S. Lorenzetti, D. Marcoccia, L. Narciso and A. Mantovani, *Reprod. Toxicol.*, 2010, **30**(1), 25–35.
53. S. Lorenzetti, I. Altieri, S. Arabi, D. Balduzzi, N. Bechi, E. Cordelli, C. Galli, F. Ietta, S. C. Modina, L. Narciso, F. Pacchierotti, P. Villani, A. Galli, G. Lazzari, A. M. Luciano, L. Paulesu, M. Spanò and A. Mantovani, *Ann. Ist. Super. Sanità*, 2011;**47**(4), 429–444.
54. V. M. Pak, L. A. McCauley and J. Pinto-Martin, *Am. Assoc. Occup. Health Nurs. J.*, 2011, **59**(5), 228–233.
55. J. J. Wirth and R. S. Mijal, *Syst. Biol. Reprod. Med.*, 2010, **56**(2), 147–167.
56. C. Byrne C, S. D. Divekar, G. B. Storchan, D. A. Parodi and M. B. Martin, *Toxicol. Appl. Pharmacol.*, 2009, **238**(3), 266–271.
57. M. Takiguchi and S. Yoshihara, *Environ. Sci.*, 2006, **13**(2), 107–116.
58. D. A. Dumesic, D. H. Abbott and V. Padmanabhan, *Rev. Endocr. Metab. Disord.*, 2007, **8**(2), 127–141.
59. J. Chen, K. C. Ahn, N. A. Gee, M. I. Ahmed, A. J. Duleba, L. Zhao, S. J. Gee, B. D. Hammock and B. L. Lasley, *Endocrinology*, 2008, **149**(3), 1173–1179.
60. Y. Xia, Y. Che, X. Zhang, C. Zhang, Y. Cao, W. Wang, P. Xu, X. Wu, L. Yi, Q. Gao and Y. Wang, *Mol. Med. Rep.*, 2012, **5**(5), 1330–1334.
61. T. Takeuchi, O. Tsutsumi, Y. Ikezuki, Y. Takai and Y. Taketani, *Endocr. J.*, 2004, **51**(2), 165–169.
62. E. Diamanti-Kandarakis and C. Piperi, *Hum. Reprod. Update*, 2005, **11**(6), 631–643.
63. J. L. San-Millán and H. F. Escobar-Morreale, *Clin. Endocrinol. (Oxf.)*, 2010, **72**(3), 383–392.
64. R. Béranger, P. Hoffmann, S. Christin-Maitre and V. Bonneterre, *Reprod. Toxicol.*, 2012, **33**(3), 269–279.
65. E. E. Hatch, R. Troisi, L. A. Wise, M. Hyer, J. R. Palmer, L. Titus-Ernstoff, W. Strohsnitter, R. Kaufman, E. Adam, K. L. Noller, A. L.

Herbst, S. Robboy, P. Hartge and R. N. Hoover, *Am. J. Epidemiol.*, 2006, **164**(7), 682–688.

66. J. A. McLachlan, R. R. Newbold, H. C. Shah, M. D. Hogan and R. L. Dixon, *Fertil. Steril.*, 1982, **38**(3), 364–371.
67. Y. C. Hu, P. H. Wang, S. Yeh, R. S. Wang, C. Xie, Q. Xu, X. Zhou, H. T. Chao, M. Y. Tsai and C. Chang, *Proc. Natl. Acad. Sci. U. S. A.*, 2004, **101**(31), 11209–11214.
68. H. Shiina, T. Matsumoto, T. Sato, K. Igarashi, J. Miyamoto, S. Takemasa, M. Sakari, I. Takada, T. Nakamura, D. Metzger, P. Chambon, J. Kanno, H. Yoshikawa and S. Kato, *Proc. Natl. Acad. Sci. U. S. A.*, 2006, **103**(1), 224–229.
69. K. A. Walters, C. M. Allan, M. Jimenez, P. R. Lim, R. A. Davey, J. D. Zajac, P. Illingworth and D. J. Handelsman, *Endocrinology*, 2007, **148**(8), 3674–3684.
70. A. Sen and S. R. Hammes, *Mol. Endocrinol.*, 2010, **24**(7), 1393–1403.
71. F. Grün and B. Blumberg, *Endocrinology*, 2006, **147**(6 Suppl.), S50–S55.
72. F. Grün and B. Blumberg, *Mol. Cell Endocrinol.*, 2009, **304**(1–2), 19–29.
73. A. Janesick and B. Blumberg, *J. Steroid Biochem. Mol. Biol.*, 2011, **127**(1–2), 4–8.
74. W. Holtcamp, *Environ. Health Perspect.*, 2012, **120**(2), a62–a68.
75. F. Grün, H. Watanabe, Z. Zamanian, L. Maeda, K. Arima, R. Chubacha, D. M. Gardiner, J. Kanno, T. Iguchi and B. Blumberg, *Mol. Endocrinol.*, 2006, **20**(9), 2141–2155.
76. F. Maranghi, S. Lorenzetti, R. Tassinari, G. Moracci, V. Tassinari, D. Marcoccia, A. Di Virgilio, A. Eusepi, A. Romeo, A. Magrelli, M. Salvatore, F. Tosto, M. Viganotti, A. Antoccia, A. Di Masi, G. Azzalin, C. Tanzarella, G. Macino, D. Taruscio and A. Mantovani, *Reprod. Toxicol.*, 2010, **9**(4), 427–432.
77. S. Lorenzetti, V. Tassinari, M. Viganotti, F. Maranghi, G. Moracci, R. Tassinari, G. Azzalin, D. Marcoccia, A. Di Virgilio, A. Eusepi, A. Romeo, A. Magrelli, M. Salvatore, F. Tosto, A. Antoccia, A. Di Masi, C. Tanzarella, G. Macino, D. Taruscio and A. Mantovani, *ISTISAN Congressi*, 11/C6, 21–22.
78. J. Wu, J. Oka, I. Tabata, M. Higuchi, T. Toda, N. Fuku, J. Ezaki, F. Sugiyama, S. Uchiyama, K. Yamada and Y. Ishimi, *J. Bone Miner. Res.*, 2006, **21**(5), 780–789.
79. M. Penza, C. Montani, A. Romani, P. Vignolini, B. Pampaloni, A. Tanini, M. L. Brandi, P. Alonso-Magdalena, A. Nadal, L. Ottobrini, O. Parolini, E. Bignotti, S. Calza, A. Maggi, P. G. Grigolato and D. Di Lorenzo, *Endocrinology*, 2006, **147**(12), 5740–5751.
80. Z. C. Dang, V. Audinot, S. E. Papapoulos, J. A. Boutin and C. W. Löwik, *J. Biol. Chem.*, 2003, **278**(2), 962–967.
81. P. Germain, B. Staels, C. Dacquet, M. Spedding and V. Laudet, *Pharmacol. Rev.*, 2006, **58**(4), 685–704.

CHAPTER 2

The Experimental 3D Structure of Nuclear Receptors. A Starting Point for Computational Investigations?

MARTIN K. SAFO*[a], GLEN E. KELLOGG[a] AND
PIETRO COZZINI[b,c]

[a] Department of Medicinal Chemistry and Institute for Structural Biology and
Drug Discovery, Virginia Commonwealth University, Richmond, VA 23298,
USA; [b] Department of Food Science, University of Parma, viale Usberti 17/A,
43100 Parma, Italy; [c] INBB, Biostructures and Biosystems National Institute,
viale Medaglie d'Oro 305, 00136 Rome, Italy
*E-mail: msafo@vcu.edu

2.1 Introduction

Nuclear receptors (NRs) define a superfamily of proteins that play major
functions in eukaryotic cell development, reproduction, differentiation and
metabolic homeostasis. The more than 150 members[1] of this family include
steroidal, non-steroidal and a multitude of orphan receptors for which their
function and native endogenous ligands remain largely unknown.[2,3] The roles
of these receptors are varied, but in general they are intracellular transcription
factors that regulate gene expression and influence reproduction, metabolism
and disposal of toxic substances.

X-ray crystallography and NMR spectroscopy and more recently computational chemistry have been used to elucidate how the different domains interact

RSC Drug Discovery Series No. 30
Computational Approaches to Nuclear Receptors
Edited by Pietro Cozzini and Glen E. Kellogg
© The Royal Society of Chemistry 2012
Published by the Royal Society of Chemistry, www.rsc.org

A

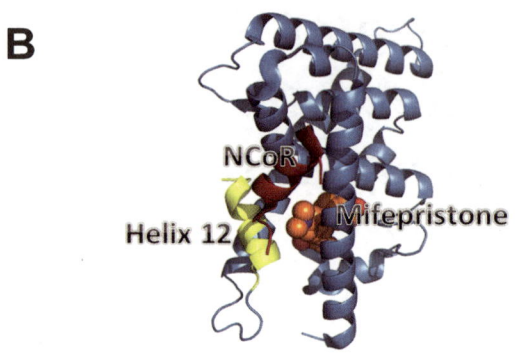

B

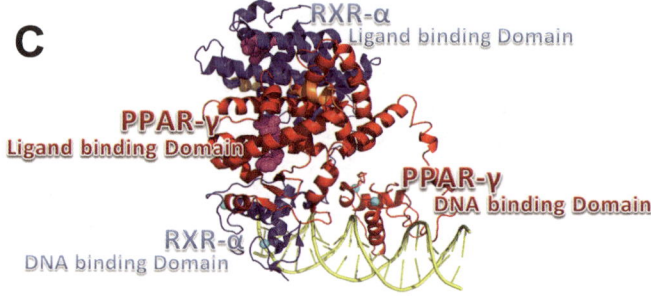

C

Figure 2.1   (a) Schematic layout of nuclear receptor structure. AF, activation function; DBD, DNA-binding domain; H, hinge; LBD, ligand-binding domain. (b) Helix 12 conformational changes and co-regulator recruitment in GR–mifepristone complex. Mifepristone displaces helix 12 from the agonist position, enlarging the co-regulator binding site and enabling the binding of NCoR (active antagonist). (c) Overall structure of the PPAR-γ–RXR-α complex on PPRE. RXR-α is blue and PPAR-γ is red. The ligands rosliglitazone and 9-*cis*-retinoic acid are shown in magenta, the Zn(II) ions are in cyan and the co-activator LXXLL peptides are in orange.

in the context of the full-length (all-atom) receptor complex and how different ligands allosterically modulate these canonical structures leading to subtle, but functionally important, changes in conformation. Currently, over 300 NR structures solved using X-ray crystallography or NMR spectroscopy have been deposited in various structural databases, including the PDB.[4] Most of these structures are for the ligand-binding domain (LBD) or DNA-binding domain (DBD), from at least one member of the NR subclasses, all showing similar folds (Figure 2.1). Also, a few structures showing more than one domain of NR have also been solved and deposited.

These structures provide unique insights into the relationship between receptor conformation and partial and full antagonism or agonism that underlie their molecular mechanism of action. The structures also suggest a window of opportunity in designing therapeutics to treat various diseases. However, an intrinsic complexity in the identification by computational methods of ligands to NRs is due to their flexibility and inherent dynamics.[5] Structural studies on the estrogen receptor (ER), which is one of the most studied NRs, indicate that selective estrogen receptor modulators (SERMs), such as tamoxifen and raloxifene, and full antagonists, such as ICI-182780 (Faslodex), bind at the same site but induce different 'open' conformations.[6,7] On the other hand, agonists, such as estrogen, stabilize the receptor 'closed' conformation.[8] The ER also shows a number of active site residues whose side chains display significantly different orientations in the various ER complexes. Other NRs are likely to have similar features and behavior. Thus, as changes in receptor conformations might have dramatic impacts on docking and other structure-based drug discovery approaches, it is crucial that protein flexibility be taken into account in understanding NR structure and, ultimately, in searching for NR ligands by computational approaches.

2.2 Flexible Proteins

2.2.1 Experimental Determination of Structure

Although the importance of flexibility is well recognized, as all proteins are inherently flexible systems – a feature often essential for their function – structure-based drug design (SBDD) has historically been generally based on static target structures.[5] Intrinsically, proteins have functionally important conformational transitions under relatively mild conditions in a wide range of time and space scales.[9–11] This conformational flexibility is particularly essential to biological function for the NRs. Most NR ligand pharmacology has been described in terms of their ability to stabilize (or move) the short H12 α-helix segment at the receptor's carboxy terminus of the receptor[12–14] and, as stated above, NR X-ray crystal structures reveal a surprisingly diverse set of ligands binding to and modulating the activities of these proteins. In gross terms, we can classify three types of proteins with respect to flexibility: (1) rigid proteins, where there are relatively small rearrangements of side chains in response to ligand binding, *etc.*, (2) flexible proteins, where the movements are relatively large, *e.g.*, at 'hinge points', active site loops, *etc.*, followed by the associated motion of the side chains, and (3) proteins whose conformaution is a function of the actual ligand binding. The majority of NRs appear to be poised within the last two, more difficult, categories. There is clearly a bias for rigid proteins in terms of those crystallographically solved (and available in databases) for a variety of reasons; one of these is the desire to reduce disorder by obtaining crystallographic data at non-biological temperatures – typically at or around 100 K. In biological systems, proteins exist in aqueous or semi-fluid environments in an ensemble of energetically accessible conformations. In this environment, the three-dimensional structure of a protein should – in some way – represent or account for all of these states. Many flexible proteins only adopt their binding site conformations in response to the pressures (induced fit) of the incoming ligand. For drug discovery in flexible targets such as the NRs, the problem is that it likely cannot be known beforehand what conformation the target will adopt when binding a particular ligand or, from the other perspective, how to design a ligand when the target conformation is not known.

High-resolution X-ray crystallography and NMR spectroscopy are the key experimental technologies for characterizing molecular structure, although a number of other methods, such as fluorescence spectroscopy,[15] spin label electron paramagnetic resonance (EPR)[16] and small-angle X-ray scattering[17] have shown promise in elucidating different aspects of protein flexibility. Importantly, over the past decade X-ray crystallography and NMR spectroscopy have been providing steadily richer and more pertinent information about the flexibility of biomacromolecules. These techniques usually only provide snapshots of one or a subset of the conformations accessible to proteins or other biomacromolecules. Generally, a high-resolution X-ray crystallographic structure is the gold standard of biological macromolecular structural models. X-ray crystallography has also been the most productive tool for structure elucidation, as the large majority of structures available in the protein data bank are from X-ray analyses. Standard and routine application of biomacromolecular crystallography produces a static, time- and space-averaged structure that mostly fails to represent accurately the dynamic nature (and structure) of a 'living' protein. Accuracy of crystal structures is usually defined by the X-ray data measurement resolution, which is a function of several factors such as the size and quality of the crystal, the brightness of the X-ray source and other experimental conditions. Resolutions of better than 2 Å are generally considered good and usable for downstream modeling, while resolutions of around 1 Å are often referred to as 'atomic'. The R-factor and free R-factor (R and R_{free}) are crystallographic measures of quality that report how well the refined structure fits the observed data.[18] Last, the B-factor, sometimes called the temperature factor, for each atom in a crystal structure represents that atom's individual uncertainty in position, which can arise from thermal motion, less than full occupancy or experimental and modeling artifacts.[19] The B-factor is an appealing metric upon which to ascribe flexibility, but the fraction of B directly related to thermal motion is uncertain and probably unknowable for protein X-ray crystal structures of moderate resolution. However, with synchrotron X-ray sources, time-resolved measurements on single crystals are now possible, as is probing the electronic oscillations around atoms to measure their probability density, either of which should reveal the inherent translational potential energy of those atoms. Also, modeling of the B-factor anisotropically,[20-22] at atomic resolution, reveals the magnitudes and directions of movement of each atom and permits a dynamic description of protein structure.

NMR spectroscopy is an alternative to X-ray crystallography that has the seemingly large advantage of being performed in solution and under conditions somewhat similar to those in the biological environment. Thus, NMR results in flexibility and dynamic motion is encoded in the reported ensembles of low-energy conformations. Each conformation is a static snapshot of the molecule; the ensemble provides a dynamic representation. For biological-scale molecules, sets of multidimensional multinuclear NMR data are analyzed in terms of the diagonal peaks that must be completely assigned and off-diagonal cross-peaks that provide information about the interaction between the protein's nuclei and the distances between their respective atoms. As hardware and software technology has advanced, NMR spectroscopy is now capable of solving protein structures in the 80–100 kDa range.[23,24]. It is now possible to solve, at high resolution, structures for proteins embedded in membranes.[25] Solution of structures from NMR data begins with a randomly folded model based on the primary sequence. The structure is optimized with either a molecular dynamics/simulated annealing protocol or with distance geometry with respect to distance and torsion angle data from experiment and known bond lengths and angles to minimize the potential energy. This generally produces the aforementioned ensemble as often more than one model satisfies the constraints. The RMSD (root-mean-square deviation) from minimum energy can be different for different regions of the protein structure, *e.g.*, flexible regions such as loops have larger deviations, as there are fewer structural constraints.[5] The highly flexible NRs, especially in their ligand-binding domains, are interesting targets for NMR structural analysis, but the intrinsic experimental difficulties of NMR analysis of larger proteins have limited its use with these targets so far. However, as the field strength of available NMR spectrometers increases with time, the size range of proteins amenable to NMR study will also increase and the resolutions of the resulting spectra will be enhanced. Hence this technique will certainly contribute to our understanding of NR structure if only because many NR proteins are difficult to crystallize owing to their flexible nature.

2.2.2   Computational Tools and Challenges

To understand and extrapolate experimental structural information on proteins further, a variety of computational approaches have been developed to aid in the solution and interpretation of the structures. In particular, molecular dynamics (MD) has become both part of the refinement protocol for X-ray crystallographic and NMR data and a key tool for the visualization and exploration of biomacromolecular structure. MD and molecular docking developed as sub-disciplines of molecular modeling, the latter being more associated with computer-aided molecular design, and in many ways these tools are complementary, especially for flexible targets.

The experimental structure models, even those from NMR data, are representative of the low-energy conformers of the molecule. To simulate higher energy conformers, MD can be performed by starting with X-ray or NMR models of the protein and subjecting these models to temperature perturbations as generated through molecular mechanics forcefields. The resulting 'movies' reveal the somewhat random motions of proteins at given temperatures and conformational sampling provides unbiased structures for further analysis. However, as MD is stochastic and simulations are usually performed on limited time scales (usually tens of nanoseconds or less), it does not guarantee that all, or even most, of the interesting conformations have been explored. MD approaches are often defined by scale, *i.e.*, the system nature and size of the molecular system, where there must be a compromise between speed and accuracy or detail. The most basic division is between coarse-grained and atom-level simulations. Coarse-grained models smooth or average out many of the atomistic details in order to investigate longer time-scale dynamics. General protein flexibility is modeled with simple normal mode analysis and Hooke's law-like potentials. Coarse-grained dynamics can provide accurate models for general protein dynamics, *e.g.*, changes resulting in the formation or reshaping of binding pockets. However, the lack of finer details often leads to poor results if these models are used as targets for drug discovery. MD simulations at the atomic level of detail simultaneously represent small atomic fluctuations and large protein movements, but the main limitation is its computational cost, which limits its universal application. Over accessible computational time frames, the extent to which the simulation samples the protein's conformational states can be questioned. Multiscale MD analysis is simply a hybrid method applying the key features of both coarse- and atomic-scale MD. As an example, consider how in the estrogen receptor α (ERα) both coarse (Figure 2.2) and atomic scales (Figure 2.3) are important in the same protein model. A last point is that the availability of increasingly inexpensive CPUs, their increasingly parallel architecture and clever algorithm development – and even the purpose-built machines of Shaw and co-workers[26] – indicates that MD will become a more and more accessible computational technology in the future.

The technology of docking and virtual screening has gone through much more evolution than MD as it relies on a somewhat more questionable concept, namely that an entire process can be understood from an examination of the endpoint, *i.e.*, docked, structure. In its earliest stages, docking involved simply placing rigid ligands into rigid sites.[27] However, this has evolved into placing flexible ligands in rigid sites[28,29] and, currently, placing flexible ligands in semi-flexible sites.[30–35] As in MD, the initial model is derived from an experimental structure. The challenge posed by the highly flexible molecules such as the nuclear receptors is modeling receptor plasticity and allowing the interacting molecules to conform to one another. This first requires knowledge of viable molecular motions; only then can a realistic docking algorithm be created. In the paradigm of high-flexibility docking, *e.g.*, where the receptor is

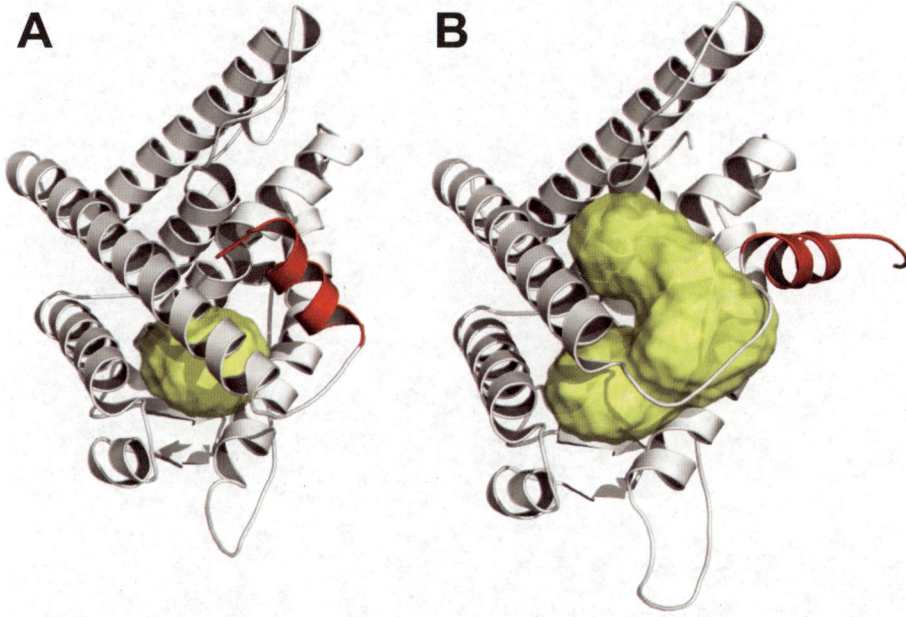

Figure 2.2 Structures of estrogen receptor α with: (a) bound agonist (17-β-estradiol; pdbid: 1G50) and (b) bound antagonist (dihydro-1,4-benzoxathiin-6-ol derivative; pdbid: 1XPC). The yellow contours indicate the volume of the ligand-binding pocket. Note also the significant shift of helix 12 (colored red).

allowed at least side-chain optimization, numerous algorithms and approaches are available. These can be categorized in a large number of ways, but for this chapter we will correlate them with two limiting case experimental observations of crystal growth. Biomacromolecular protein–ligand complex crystals suitable for X-ray analysis are grown either by co-crystallization of the protein and ligand or by soaking the ligand into a preformed crystal of the unliganded protein. In the former case, one can imagine the protein undergoing functionally relevant conformational motion while the ligand selects (by binding to) the proper and compatible protein conformation from those available.[9–11] This represents the induced fit approach, whereas the active site rationally adapts to the incoming ligand with side-chain motions that are known to occur[36–39] while taking advantage of low-energy conformational changes. However, with the soaking method, which only works in some cases, the preformed static lattices appear to select ligands with conformations that fit the available site. One possible negative result of attempting to soak a ligand into a crystal is that the crystal shatters, as the required adaptation is more than the lattice can handle. Docking applications that correlate with the soaking methodology are those that allow the ligand to adapt multiple conformations in the binding site with only limited side-chain motion for the surrounding residues, *e.g.*, GOLD.[40]

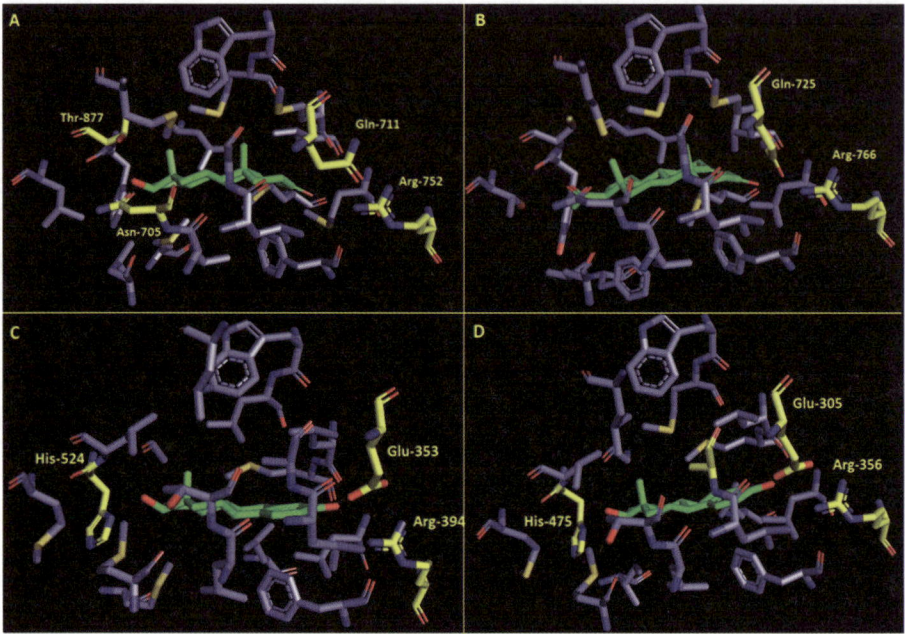

Figure 2.3 Depictions of the ligand-binding domain (LBD) for nuclear receptors that show local flexibility. (a) Androgen receptor (pdbid: 2AM9) complexed with testosterone; (b) progesterone receptor (1A28) complexed with progesterone; (c) estrogen receptor α (2YJA) complexed with 17β-estradiol; and (d) estrogen receptor β (3OLL) complexed with 17β-estradiol. Residues belonging to side chains near the ligands are shown, colored as follows: blue for amino acids involved in hydrophobic interactions, yellow for those involved in hydrogen bonds; the ligands are colored green. His524 in ERα and His475 in ERβ are most responsible for local flexibility.

2.3 Nuclear Receptors

2.3.1 A Closer Look at Structure and Function

The NR superfamily is typically divided into the steroid receptor family, which includes the estrogen receptor (ER), progesterone receptor (PR), glucocorticoid receptor (GR), androgen receptor (AR) and mineralocorticoid receptor. The thyroid or retinoid family (class II) includes the thyroid receptor (TR), vitamin D receptor (VDR), retinoic acid receptor (RAR) and the peroxisome proliferator receptor (PPAR). The third class, which is often termed the orphan receptor family, defines other receptors that currently have no known or identified cognate ligand.

NRs exert their functions by ligand activation and they usually contains six functional domains (A–F), including the first transcription activation domain (AF-1, A/B), DNA-binding domain (C, also known as the DBD), hinge domain (D) and the second transcription activation domain (AF-2, E/F), also referred to as the ligand-binding domain (LBD) (*e.g.*, see Figure 2.1c). The AF-2 domain is highly regulated by compounds that bind to the LBD. Importantly, molecular interactions between ligands and the LBD, and also the inherent conformational changes, are central to regulation of gene expression by the nuclear hormone receptors and are essential for proper control of the cell cycle and differentiation. The LBDs of different NRs thus have unique binding pockets specific for their cognate hormone or ligand. Metabolites and xenobiotics can also bind to the LBD. The LBD sequence varies substantially between NRs, but the structure is well conserved with 11–13 α-helices organized around a hydrophobic binding pocket. Residues within the binding pocket confer specificity, determining whether the LBD will accept steroid hormones or retinoid compounds or bind xenobiotic ligands that affect receptor function. LBDs also contain nuclear localization signals, interaction motifs for heat shock proteins, coregulators and other transcription factors. In addition to its ligand-dependent properties, the AF-2 region of the LBD serves to recruit various co-activating proteins that are required for the transcriptional activity of the protein. The LBD is also involved in oligomerization necessary for high-affinity DNA response element binding to the DBD.

The DBD is involved in response element binding and also serves as allosteric transmitter of information to other regions of the receptor protein. It is also the most highly conserved domain among the members of the NR superfamily, consisting of two zinc-finger motifs. The tertiary structure of the DBD contains helices that bind to specific sequences of DNA called hormone response elements.

The DNA- and ligand-binding domains are connected by the short hinge region D and the specific function of this domain is unclear. N-Terminal sequence beyond the DBD may be found in several NRs containing the first transcriptional domain (AF-1). Unlike the LBD and DBD, which a have high degree of homology among the NRs, the AF-1 domain shows weak conservation.[41] It can function as a ligand-independent transcriptional activator, and also act synergistically with AF-2.[42]

2.3.2 Survey of Current Experimental Structural Data

Both experimental, X-ray crystallographic and NMR and computational techniques have been used to explore the interactions between domains in the full-length receptor complex and between the ligands and receptors. Although more than 300 NR structures are available, most of these structures are for the LBD or DBD; see Nwachukwu and Nettles,[43] Kumar and Thompson,[44] Bain *et al.*[45] and Jin and Li[46] for reviews. Unlike the DBDs and LBD, very little structural information is available for the N-terminal transcriptional AF-1 domain (NTD). The NTD is flexible and studies with circular dichroism and NMR spectroscopy suggest this domain to be largely unstructured, but it becomes structured, possibly involving α-helix structure, in response to protein–protein interactions and structure-stabilizing solutes.[44,47] The extreme C-terminus of NR, referred to as the F domain, is understudied not only functionally, but also structurally. In a study with the ER, it was suggested that the F domain contributes to differences in the activities of ER subtypes (α and β subtypes), and that this domain was required for tamoxifen's agonist activity on an estrogen response element, by modifying the receptor's interactions with coregulators.[48] The differences between the F domains of the ER α and β subtypes and among the other members of the nuclear hormone receptor superfamily may offer opportunities for selective control of the activity of these proteins.[48]

A few structures showing more than one domain of NR are also known.[49] The crystal structure of the retinoid X α receptor/PPARγ complex has been solved using X-ray crystallography.[6] The structure includes the DBD, LBD and hinge region of PPARγ/RXRα bound with DNA, thus providing important molecular bases for understanding domain organization and target gene recognition. The A/B regions are absent from the crystal structure, which was suggested to be due to flexibility. In a recent study, cryo-electron microscopy was used to determine the complex structure of the vitamin D receptor (VDR) with RXR and their cognate DNA response element.[50] The architectures of the two full-length structures were found to be significantly different from each other, with the PPARγ/RXR showing a more compact structure and close conformation. Interestingly, recent small-angle X-ray scattering experiments showed that both RAR/RXR and PPARγ/RXR adopt open extended conformations in solution.[51] These studies underscore the importance of the hinge domains that allow for conformational flexibility. Nonetheless, it was noted that the single long helix in the hinge of VDR would interfere with adoption of a compact architecture, making its complex more rigid than PPARγ/RXR and other NRs.[43]

Surprisingly, whereas NMR spectroscopy has played a significant historic role in elucidating the structures of DBDs from various NRs, recent NMR investigations on these flexible proteins are relatively rare. Structural comparison of several DBDs with their counterparts derived from NMR spectroscopy generally revealed close agreement, particularly for ER and GR subunits.[52–56] Whereas the first zinc-finger of the DBD is well resolved and conformationally similar for the most part in both the NMR and X-ray structures, the second zinc-finger region of the DBD as solved by NMR spectroscopy is poorly defined compared with the well-resolved region in the X-ray crystal structure.[55] Part of this second zinc-finger region is known to form a distorted helix as it undergoes a conformational change upon cooperative binding to DNA. NMR relaxation measurements and MD simulations of the GR DBD suggest a uniform and limited mobility along the backbone.[57,58] Nonetheless, concerted motions in and between the subdomains have been proposed to facilitate structural rearrangements that contribute to the cooperativity of DNA binding.[57,58] Differences between the unbound GR DBD determined using NMR spectroscopy and the DNA-bound DBD determined using X-ray analysis have been reported.[59] There is apparently a distortion in one of the α-helices (residues 42–45) when DNA binds, and this is necessary for interaction between the DNA and the protein. Similar differences between NMR and X-ray crystallographic structures have been observed for the ER DBD.[55,59] These experimental structures provide unique insights into the relationship between receptor conformation and partial and full antagonism or agonism that underlie their molecular mechanism of action. Significantly, in the best cases, the structures also offer a window of opportunity in designing therapeutics to treat several diseases.

Elucidating the dynamic structure of NRs, or their energetically accessible alternative conformations, has mostly relied on MD simulations that use as a starting point experimental structural data for the NR or, with less certainty, homology models of the NR built from related proteins. There have been a quite a few recent MD studies of note; we mention a few here. Teotico *et al.*[60] showed, in 20–30 ns simulations, that NRs [pregnane X receptor (PXR), PPAR-γ and ER-α in this study] in their active-ready conformations show correlated AF-2 domain motions. This suggests that these nuclear receptors possess a preformed protein–protein interaction surface that is consistently ready for co-activator contacts. Souza *et al.* examined three mutations at Ile280 of H12 of thyroid hormone receptors with mutagenesis, assays and MD.[61] Interestingly, this residue does not interact with either co-activators or ligand, yet its mutation can inhibit co-activator/co-repressor binding and differentially affect the receptor's binding affinity for the hormone. The simulations indicate that the mutations indirectly displace the ligand from the binding pocket, with a secondary effect of allowing water penetration to destabilize further the ligand's binding. Two mutations (to Arg and Lys) allow the formation of a salt bridge between H12 and the LBD, which blocks recruitment by stabilizing H12. The third mutation (to Met) blocks co-repressor binding and enhances co-activator affinity, which suggests that H12 has been locked in an agonist conformation. In a similar manner, Elhaji *et al.* showed with MD that *in silico* mutations of the conserved Pro892 can be correlated with clinical observations of individuals with complete androgen insensitivity syndrome (with mutations to either Ala or Leu).[62] The Ala-mutated AR dynamics showed reduced ligand binding and transactivational potential due to increased H12, while the dynamics of the Leu-mutated receptor produced a structure with H12 out of position and distorted. The role of the Pro at this position in other NRs may be similarly critical.

2.3.3 Mechanism of Action: Insight from Ligand Binding

Small-molecule ligands play important roles in modulating the activity of NR, since the binding of ligands can induce the conformational changes that determine the recruitment of co-activators or co-repressors. Thus, the specific functions of NR are tightly associated with their cognate ligands. As indicated above, the nuclear receptors are highly dynamic, in that binding of a ligand,

DNA or transcriptional co-regulator proteins can allosterically change the conformation of the protein, affecting subsequent binding events. It has been proposed that NRs be viewed as multi-modal scaffold proteins,[43] where binding of a ligand or DNA can alter the nature of the scaffold and thus be a determinant for subsequent molecular interactions and activity profiles. The crystal structures of the LBD of several NRs, and also their co-complexes bound with agonists and antagonists and co-regulators, show insight into the agonistic/antagonistic behavior of NRs. The NR LBD structures show similar globular conformation that is made up of about 12 helices. In most of these structures, the ligand-binding pocket is formed by four of the helices. The binding pockets from different NRs are very unique in terms of size, geometry and hydrophobic and hydrophilic residues, allowing these receptors to discriminate among closely related ligands. For example, steroidal receptors, such as AR and ER that bind to relatively few ligands, have small binding pockets with several hydrophilic residues. In contrast, the orphan receptors which interact with diverse metabolic ligands tend to have larger binding pockets.

Co-activator/co-repressor recruitment by the receptors is primarily mediated by the AF-2 domain, which is located in the LBD between helices H3, H5 and H12. Structural analysis shows that AF-2 activity is dictated by the orientation of the mobile helix H12. Agonist binding stabilizes an 'active' orientation of H12 resulting in the formation of a specific binding site for the NR co-activators. NR antagonists sterically prevent helix H12 from adopting the agonist-induced conformation and may result in the AF-2 binding site being accessible to interact with the NR co-repressors. For instance, estrogen binding to the ER LBD induces a shift in helix H12, which closes the LBD pocket and provides a surface to which co-activators can bind. In contrast, 4-hydroxytamoxifen and raloxifene, which possess bulky amine side chains (anti-estrogenic side chains), prevent the proper positioning of helix H12 and consequently destabilizes the ER–co-activator interaction surface, preventing the formation of the co-activator recruitment surface, and impede the ultimate gene expression.[63] Similarly, binding of asoprisnil to PR also shows a displacement of helix H12 from the agonistic position to the antagonistic position.[64] Several structures of the GR ligand-binding domain (GR-LBD) with mifepristone show that the ligand triggers the helix H12 molecular switch to reshape the co-activator site into the co-repressor site [65]. In some NRs, such as RXR, oligomerization of the proteins also physically prevents co-activation binding. Agonist binding is necessary for dissociation of the tetramers, allowing recruitment of co-activators.

In a recent study by Zhakarov et al.[66] with AR, it was noted that co-activator association resulted in conformational rearrangement in the AR-LBD and different co-activators induce distinct conformational states in the AR complex. It was also reported that binding of the co-activator induced specific alterations in the backbone flexibility of the AR-LBD distant from the site of binding. These data suggest that, even in the presence of same ligand,

NR-LBDs can occupy distinct conformational states depending on the interactions with specific co-activators in the tissues. In a computational study with the vitamin D receptor (VDR) in a complex with its natural ligand and an analog thereof, no detectable changes in the backbone configuration or ligand topology in the receptor-binding cavity were reported, although binding of the ligand led to a dramatic increase in functional activity.[67] Nonetheless, the authors observed significant changes in the side-chain orientations of more distant amino acids that are known to be involved in co-activator interactions and/or involved in VDR–RXR heterodimerization.

Computational studies using molecular dynamics have attempted to illuminate the ligand-binding process by watching the unbinding process with steered MD, targeted MD or random acceleration MD. These three techniques were used together to study the pathways for unbinding 1α,25-dihydroxyvitamin D_3 from the LBD of the VDR.[68] These simulations suggest that an exit pathway located between the H1–H2 loop and the β-sheet between the H5 and H6 helices is favored over a pathway found between H3 and the H1–H2 loop. Neither path appears to require a displacement of H12. In contrast, one of the three paths observed for dissociating 17β-estradiol from the ER with locally enhanced sampling MD molecular dynamics does involve repositioning of H12, whereas both of the others proceed by separating H8 and H11.[69] However, dimerization of ER appears to close down these last two paths. This observation suggests that ligand dissociation rates may be at least partly regulated by this dimerization, perhaps throughout the NR family.

2.4 Issues for Structure-based Drug Design in Nuclear Receptors

Research in the last decade has provided significant insight into why certain ligands act as antagonists in some cells and agonists in others, or why, in the same cell, various structurally related ligands show agonist or antagonist activity. These ligands are known to bind to NRs causing subtle but critical changes in the shape of their receptors, resulting in conformational changes that lead to differing patterns of interactions with co-activators/co-repressors/other transcription factors. The major challenge of NR-based drug discovery is thus how to design either pure agonists or pure antagonists for their respective beneficial effects while avoiding other undesirable effects. Most attempts to overcome this challenge have been blocked by the lack of experimental structural data, from either NMR spectroscopy or X-ray crystallography, which would light the way to rational drug design. We remain unable *really* to access and understand the distinct conformational states that are linked to ligand binding, and also map the interactions that are unique for either agonistic or antagonistic behavior of the receptor. The lack of experimental data is, in turn, largely due to the particularly flexible nature of the nuclear receptor structures. Nature has also designed these proteins to be flexible in their roles and functions. Several LBD structures have been solved in the presence of synthetic ligands and these have revealed unique positions and functional details of the helix H12 molecular switch that can be related to active and passive antagonism, and also agonism and partial agonism.[65] Computational modeling has also been used successfully to show that even in the presence of the same ligand, the LBD can occupy distinct conformational states depending on its interactions with specific co-activators in tissue, explaining why the same ligand can act as an agonist and/or antagonist in different cells or tissues.[70]

Nevertheless, there is a highly significant body of work that can be accomplished by leveraging the existing experimental data with computational tools.[71] These methods can be applied in a variety of ways to probe important receptor–ligand and protein–protein interactions. The resulting understanding has led to new, potent and specific chemical agents that are agonists and antagonists and provides a window into off-target interactions that may lead to unwanted side effects and toxicity. The most relevant approach to dock potential ligands into sites that are as flexible as those of the NRs is to use induced fit docking. Two recent reviews[72,73] have explored this topic. It should be clear by now that the malleability of these binding pockets makes it difficult to extrapolate the knowledge gained from one crystal structure of a particular receptor–ligand complex to the binding of a different ligand with that same receptor target, often because of the macroscopic flexibility of H12 and also the expected local flexibility of side chains in the pocket. Thus, MD and induced fit docking, despite their limitations, are crucial tools for understanding and exploiting these receptors as drug targets. As these tools become more available, are made more efficient with highly parallel computer architectures and are primed with better and more accurate experimental structural data, the ability to discover and/or design new and selective ligands will be enhanced in the future. The remaining chapters in this book describe several diverse perspectives of understanding NRs by both experimental and theoretical means. The numerous health benefits and health risks of small-molecule binding to NRs suggest that it is imperative to continue these studies.

References

1. D. J. Mangelsdorf, C. Thummel, M. Beato, P. Herrlich, G. Schütz, K. Umesono, B. Blumberg, P. Kastner, M. Mark, P. Chambon and R. M. Evans, *Cell*, 1995, **83**, 835.
2. P. Ciana, E. Vegeto, M. Beato, P. Chambon, J. Å. Gustafsson, M. Parker, W. Wahli and A. Maggi, *EMBO Rep.*, 2002, **3**, 125.
3. V. Giguère, *Endocr. Rev.*, 1999, **20**, 689.
4. I. J. McEwan and A. M. Nardulli, *Nucl. Recept. Signal.*, 2009, **7**, e011.
5. P. Cozzini, G. E. Kellogg, F. Spyrakis, D. J. Abraham, G. Costantino, A. Emerson, F. Fanelli, H. Gohlke, L. A. Kuhn, G. M. Morris, M. Orozco, T. A. Pertinhez, M. Rizzi and C. A. Sotriffer, *J. Med. Chem.*, 2008, **51**, 6237.

6. V. Chandra, P. Huang, Y. Hamuro, S. Raghuram, Y. Wang, T. P. Burris and F. Rastinejad, *Nature*, 2008, **456**, 350.

7. A. C. W. Pike, A. M. Brzozowski, J. Walton, R. E. Hubbard, A. G. Thorsell, Y. L. Li, J.-A. Gustafsson and M. Carlquist, *Structure*, 2001, **9**, 145.

8. B. Sandak, H. J. Wolfson and R. Nussinov, *Proteins: Struct. Funct. Genet.*, 1998, **32**, 159.

9. C. J. Tsai, S. Kumar, B. Ma and R. Nussinov, *Protein Sci.*, 1999, **8**, 1181.

10. I. Bahar, C. Chennubhotla and D. Tobi, *Curr. Opin. Struct. Biol.*, 2007, **17**, 633.

11. K. Henzler-Wildman and D. Kern, *Nature*, 2007, **450**, 964.

12. G. Krauss, *Biochemistry of Signal Transduction and Regulation*, Wiley-VCH, Weinheim, 2003, p. 541.

13. C. Hellal-Levy, J. Fagart, A. Souque, J.-M. Wurtz, D. Moras and M.-E. Rafestin-Oblin, *Mol. Endocrinol.*, 2000, **14**, 1210.

14. X. Hu and M. A. Lazar, *Trends Endocrinol. Metab.*, 2000, **11**, 6.

15. B. Somogyi, Z. Lakos, A. Szarka and M. Nyitrai, *J. Photochem. Photobiol. B*, 2000, **59**, 26.

16. W. L. Hubbell, D. S. Cafiso and C. Altenbach, *Nat. Struct. Biol.*, 2000, **7**, 735.

17. J. Lipfert and S. Doniach, *Annu. Rev. Biophys. Biomol. Struct.*, 2007, **36**, 307.

18. L. Jensen, *Methods Enzymol.*, 1997, **B277**, 353.

19. J. Dunitz, V. Shomaker and K. Trueblood, *J. Phys. Chem.*, 1988, **92**, 856.

20. E. A. Merritt, *Acta Crystallogr. D Biol. Crystallogr.*, 1999, **55**, 1109.

21. D. Vitkup, D. Ringe, M. Karplus and G. A. Petsko, *Proteins*, 2002, **46**, 345.

22. A. Schmidt and V. S. Lamzin, *Cell Mol. Life Sci.*, 2007, **64**, 1959.

23. R. Horst, G. Wider, J. Fiaux, E. B. Bertelsen, A. L. Horwich and K. Wüthrich, *Proc. Natl. Acad. Sci. U. S. A.*, 2006, **103**, 15445.

24. A. Grishaev, V. Tugarinov, L. E. Kay, J. Trewhella and A. Bax, *J. Biomol. NMR*, 2008, **40**, 95.

25. B. Liang and L. K. Tamm, *Proc. Natl. Acad. Sci. U. S. A.*, 2007, **104**, 16140.

26. K. Lindorff-Larsen, N. Trbovic, P. Maragakis, S. Piana and D. E. Shaw, *J. Am. Chem. Soc.*, 2012, **136**, 3787.

27. I. D. Kuntz, J. M. Blaney, S. J. Oatley, R. Langridge and T. E. Ferrin, *J. Mol. Biol.*, 1982, **161**, 269.

28. M. Rarey, B. Kramer, T. Lengauer and G. A. Klebe, *J. Mol. Biol.*, 1996, **261**, 470.

29. G. M. Morris, D. S. Goodsell, R. S. Halliday, R. Huey, W. E. Hart, R. K. Belew and A. J. Olson, *J. Comput. Chem.*, 1998, **19**, 1639.

30. H. Claussen, C. Buning, M. Rarey and T. Lengauer, *J. Mol. Biol.*, 2001, **308**, 377.

31. C. N. Cavasotto and R. A. Abagyan, *J. Mol. Biol.*, 2004, **337**, 209.

32. F. Osterberg, G. M. Morris, M. F. Sanner, A. J. Olson and D. S. Goodsell, *Proteins: Struct. Funct. Genet.*, 2002, **46**, 34.

33. A. N. Jain, *J. Comput.-Aided Mol. Des.*, 2007, **21**, 281.

34. M. L. Verdonk, J. C. Cole, M. J. Hartshorn, C. W. Murray and R. D. Taylor, *Proteins: Struct. Funct. Genet.*, 2003, **52**, 609–623.

35. B. Q. Wei, L. H. Weaver, A. M. Ferrari, B. W. Matthews and B. K. Shoichet, *J. Mol. Biol.*, 2004, **337**, 1161.

36. R. Najmanovich, J. Kuttner, V. Sobolev and M. Edelman, *Proteins: Struct. Funct. Genet.*, 2000, **39**, 261.

37. M. I. Zavodszky and L. A. Kuhn, *Protein Sci.*, 2005, **14**, 1104.

38. K. Gunasekaran and R. Nussinov, *J. Mol. Biol.*, 2007, **365**, 257.

39. A. Gutteridge and J. Thornton, *J. Mol. Biol.*, 2005, **346**, 21.

40. G. Jones, P. Willett and R. C. Glen, *J. Mol. Biol.*, 1995, **245**, 43.

41. A. Wärnmark, E. Treuter, A. P. Wright and J. A. Gustafsson, *Mol. Endocrinol.*, 2003, **17**, 1901.

42. G. A Francis, E. Fayard, F. Picard and J. Auwerx, *Annu. Rev. Physiol.*, 2003, **65**, 261.

43. J. C. Nwachukwu and K. W. Nettles, *EMBO J.*, 2012, **31**, 251.

44. R. Kumar and E. B. Thompson, *Steroids*, 1999, **64**, 310.

45. D. L. Bain, A. F. Heneghan, K. D. Connaghan-Jones and M. T. Miura, *Annu. Rev. Physiol.*, 2007, **69**, 201.

46. L. Jin, Y. Li, *Adv. Drug Deliv. Rev.*, 2010, **62**, 1218.

47. D. N. Lavery and I. J. McEwan, *Biochem. J.*, 2005, **391**, 449.

48. D. F. Skafar and C. Zhao, *Endocrine*, 2008, **33**, 1.

49. F. Rastinejad, T. Perlmann, R. M. Evans and P. B. Sigler, *Nature*, 1995, **375**, 203.

50. I. Orlov, N. Rochel, D. Moras and B. P. Klaholz, *EMBO J.*, 2012, **31**, 291.

51. N. Rochel, F. Ciesielski, J. Godet, E. Moman, M. Roessle, C. Peluso-Iltis, M. Moulin, M. Haertlein, P. Callow, Y. Mely, D. I. Svergun and D. Moras, *Nat. Struct. Mol. Biol.*, 2011, **18**, 564.

52. T. Hard, E. Kellenbach, R. Boelens, B. A. Maler, K. Dahlman, L. P. Freedman, J. Carlstedt-Duke, K. R. Yamamoto, J. A. Gustafsson and R. Kaptein, *Science*, 1990, **249**, 157.

53. H. Baumann, K. Paulsen, H. Kovacs, H. Berglund, A. P. H. Wright, J. A. Gustafsson and T. Hard, *Biochemistry*, 1993, **32**, 13463.

54. J. W. R. Schwabe, L. Chapman, J. T. Finch and D. Rhodes, *Cell* 1993, **75**, 567.

55. J. W. R. Schwabe, D. Neuhaus and D. Rhodes, *Nature*, 1990, **348**, 458.

56. B. F. Luisi, W. X. Xu, Z. Otwinowski, L. P. Freedman, K. R. Yamamoto and P. B. Sigler, *Nature*, 1991, **352**, 497.

57. M. A. L. Eriksson, H. Berglund, T. Hard and L. Nilson, *Proteins Struct. Funct. Genet.*, 1993, **17**, 375.

58. H. Berglund, H. Kovacs, K. Dahlman-Wright, J. A. Gustafsson and T. Hard, *Biochemistry*, 1992, **31**, 12001.

59. M. A. A. van Tilborg, A. M. J. J. Bonvin, K. Hård, A. L. Davis, B. Maler, R. Boelens, K. R. Yamamoto and R. Kaptein, *J. Mol. Biol.*, 1997, **247**, 689.

60. D. G. Teotico, M. L. Frazier F. Ding, N. V. Dokholyan, B. R. S. Temple and M. R. Redinbo, *PLoS Comput. Biol.*, 2008, **4**, e1000111.

61. P. C. T. Souza, G. B. Barra, L. F. R. Velasco, I. C. J. Ribeiro, L. A. Simeoni, M. Togashi, P. Webb, F. A. R. Neves, M. S. Skaf, L. Martínez and I. Polikarpov, *J. Mol. Biol.*, 2011, **412**, 882.

62. Y. A. Elhaji, I. Stoica, S. Dennis, E. O. Purisima and M. A. Trifiro, *Hum. Mol. Genet.*, 2006, **15**, 921.

63. A. M. Brzozowski, A. C. W. Pike, D. Zbigniew, R. E. Hubbard, T. Bonn, O. Engstrom, L. Ohman, G. L. Greene, J.-A. Gustafsson and M. Carlquist, *Nature*, 1997, **389**, 753.

64. K. P. Madauss, E. T. Grygielko, S. J. Deng, A. C. Sulpizio, T. B. Stanley, C. Wu, S. A. Short, S. K. Thompson, E. L. Stewart, N. J. Laping, S. P. Williams and J. D. Bray, *Mol. Endocrinol.*, 2007, **21**, 1066.

65. G. A. Schoch, B. D'Arcy, M. Stihle, D. Burger, D. Bär, J. Benz, R. Thoma and A. Ruf, *J. Mol Biol.*, 2010, **395**, 568.

66. M. N. Zakharov, B. K. Pillai, S. Bhasin, J. Ulloor, A. Y. Istomin, C. Guo, A. Godzik, R. Kumar and R. Jasuja, *Mol. Cell. Endocrinol.*, 2011, **341**, 1.

67. W. Sicinska and P. Rotkiewicz, *J. Steroid Biochem. Mol. Biol.*, 2009, **113**, 253.

68. M. Peräkylä, *Eur. Biophys. J.*, 2009, **38**, 185.

69. M. T. Sonoda, L. Martínez, P. Webb, M. S. Skaf and I. Polikarpov, *Mol. Endocrinol.*, 2008, **22**, 1565.

70. L. Bergman, M. L. Beelen, M. P. Gallee, H. Hollema, J. Benraadt and F. E. Van-Leeuwen, *Lancet*, 2000, **356**, 881.

71. N. Ai, M. D. Krasowski, W. J. Welsh and S. Ekins, *Drug Discov. Today*, 2009, **14**, 486.

72. W. Cornell and K. Nam, *Curr. Top. Med. Chem.*, 2009, **9**, 844.

73. M. Nocker and P. Cozzini, *Curr. Top. Med. Chem.*, 2011, **11**, 133.

CHAPTER 3

Protein Structure Analysis with Constraint Programming

ALESSANDRO DAL PALÙ*[a], AGOSTINO DOVIER[b], FEDERICO FOGOLARI[c] AND ENRICO PONTELLI[d]

[a] Dipartimento di Matematica e Informatica, Università di Parma, Viale Usberti 53/A, I-43100 Parma Italy; [b] Dipartimento di Matematica e Informatica, Università di Udine, Via delle Scienze 206, 33100 UDINE Italy; [c] Dipartimento di Scienze Mediche e Biologiche, Università di Udine, Piazzale Kolbe, 4, 33100 UDINE Italy; [d] Department of Computer Science, New Mexico State University, Box 30001 MSC CS, 88003-8001 Las Cruces, NM, USA
*E-mail: alessandro.dalpalu@unipr.it

3.1 Introduction

In this chapter, we present a review of various approaches to modeling protein structure by means of constraint programming.[1] This particular programming framework has been used in the modeling of the structure and properties of chemical compounds. Even though not as widespread as other approaches (*e.g.*, those based on physical and chemical simulations), this methodology offers several advantages in terms of declarative modeling and search of solutions. For example, it is possible (1) to describe the position and mobility of objects through *constraints*, (2) to employ *reasoning* techniques to reduce the set of potential solutions and (3) to use energetic or cost constraints in order to formulate an optimization problem.

After a brief introduction to constraint programming and some notes about its use with molecular structures, we present various applications ranging from

RSC Drug Discovery Series No. 30
Computational Approaches to Nuclear Receptors
Edited by Pietro Cozzini and Glen E. Kellogg
© The Royal Society of Chemistry 2012
Published by the Royal Society of Chemistry, www.rsc.org

proof of concepts to working packages (see, *e.g.*, PSICO[2]) based on this technology.

3.2 A Constraint Programming Primer

Constraint programming (CP)[1,3] originates from a branch of artificial intelligence and, in particular, from the study of combinatorial and optimization problem solving. CP allows one to tackle a wide variety of problems that traditionally arise in areas such as job scheduling, timetabling, routing and industrial planning. The key feature of this programming methodology is that the focus is on the definition and handling of general properties (*constraints*) of the model being studied. This methodology is able to model complex problems and to solve them efficiently.

Problem modeling is the most delicate phase, since it determines the features of the problem that are enabled to interact through constraints. The features of the problem are abstracted using *variables*, that can range in a specific *domain*, *e.g.*, integers or real numbers, and using algebraic operations and relations defined over such domains. Solutions to a problem will be captured by assignment of values to the variables, where the values are constrained to be drawn from the specified domains.

A CP model contains mathematical expressions or relations (*constraints*), *e.g.*, $X > 0$, $Y + Z < 1$, that have to be satisfied by every solution. The process of searching for solutions that satisfy every constraint is referred to as *constraint solving* and it is performed by specialized algorithms, called *constraint solvers*. Constraint solvers are specialized to work on specific domains; the most common ones can work on domains containing integer numbers, real numbers and sets.[4]

The core of the constraint programming methodology is based on a *constrain and generate* technique. The *constrain* phase formulates the constraints that have to be satisfied, restricting the space of solutions, whereas the *generate* phase searches for the compatible assignments to variables that satisfy all the constraints.

In general, the time needed to explore the solution space can grow beyond any available computational power; therefore, specific techniques (called constraint *propagation* and *filtering*) are employed during the search to reduce drastically the time spent on the generation of the solutions. The key idea is to delay as much as possible the assignments to variables, while using reasoning techniques to remove some of the values in the domains that cannot be feasibly used in any solutions. These techniques rely on the analysis of the available constraints and of the domains of the variables and can lead to drastic reductions in the search space to be explored. Constraint resolution techniques come from different areas of mathematics and computer science, *e.g.*, operational research and linear programming.

The interaction of constraints with the search space has a cost that has to be considered. In general, the cost of constraint handling (*e.g.*, through propagation and filtering) is greatly surpassed by the significant savings in time spent searching the admissible fraction of the search space.

Let us present here the basic structure of the approach:

$$\text{Find}(X):$$

$$\text{Constrain}(X),$$

$$\text{Generate}(X).$$

From the development point of view, a satisfactory model is often the result of several refinement steps, where constraints are edited and tested.

Since constraints are independent of one another and have a modular nature, it is possible to add and edit them during the development phase with no modifications to the constraint solver. This is a crucial property that allows fast prototyping and testing before applying the model to larger and/or more refined instances of the problem. More traditional programming frameworks often embed the model properties and constraints into resolution strategies, by giving rise to a monolithic program. As a consequence, any change to the model requires non-trivial modifications to the resolution algorithm. The clear separation between problem modeling and problem solving allows one to exploit constraints and to perform some reasoning about them.

A popular example used to illustrate the CP model is the word-for-number problem of assigning different numbers from 0 to 9 to each letter so that the following expression holds:

$$\text{SEND} + \text{MORE} = \text{MONEY}$$

Even if the problem has a large number of possible assignments, it can be solved by anyone in a few minutes, with the incremental application of constraints while reasoning.

Some CP languages offer direct control of the search strategy, approximated searches and integrations of active research topics in local search,[5] such as hill-climbing, simulated annealing and genetic algorithms.

In the following sections, we provide more details of problem formalization and resolution.

3.2.1   Constraint Satisfaction Problems

Let us briefly define the main notions of constraint programming. Let X_1, ..., X_k be a list of *variables*. Every variable X_i is associated with a set of possible values D_i, called its *domain*. The most common domains are expressed as subsets of the integer (Z) or real (R) numbers, often defined as one-dimensional (1D) intervals (*e.g.*, 1...8 denotes the set of integer numbers {1, 2, 3, 4, 5, 6, 7, 8}; [1,8] denotes the set of real numbers between 1 and 8). More complex domains are also allowed. Those of interest for biochemical

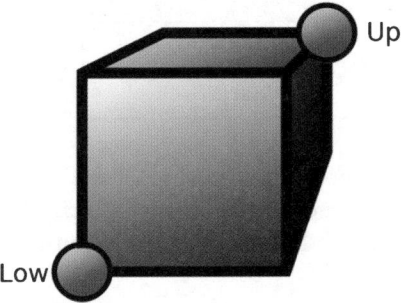

Figure 3.1   A 3D domain.

applications are the sets of three-dimensional (3D) points that can be used for representing the spatial positions of atoms. A compact method to represent a set of 3D points is the extension in 3D of the 1D interval, *i.e.*, a box identified by two opposite corners Low and Up (Figure 3.1).

A *constraint* over some variables $X_1, ..., X_n$ is a relation over their domains $D_1, ..., D_n$. Syntactically, constraints are expressed by simple arithmetic functions and relations such as $2X_1 + 3X_2X_3 < 6X_4$, $X_1 + X_2 = 0$.

A *constraint satisfaction problem* (CSP) consists of a list $X_1, ..., X_k$ of variables with corresponding domains $D_1, ..., D_k$ and a set C of constraints among subsets of the variables. A *solution* of the CSP is an assignment $X_1 = d_1, ..., X_k = d_k$, where d_i belongs to D_i and each constraint C is satisfied by the assignment. If a CSP admits a solution, then it will be said to be *consistent* or satisfiable, otherwise it is not consistent.

Example

Let us consider again the SEND + MORE = MONEY example. It can be modeled as a CSP in the following way. The list of variables is S, E, N, D, M, O, R, Y; each variable has the domain 0...9. We can impose a number of constraints to model the arithmetic properties of the problem. The variables S and M, representing the most significant digits of the three numbers, may not be equal to zero; therefore, the constraints $S \neq 0$ and $M \neq 0$ are added.

Moreover, the puzzle requires that different variables are assigned to different numbers. This can be enforced by a list of constraints of the kind $S \neq E$, $S \neq N$, ..., $R \neq Y$. A common built-in constraint provided by most constraint solvers is *alldifferent(list of variables)*, used to indicate that all the variables in the list must be assigned pairwise distinct values. Thus, in our example, we can introduce the constraint *alldifferent(S,E,N,D,M,O,R,Y)*. (Being useful in many problems, the *alldifferent* constraint has been deeply studied. In particular, extremely efficient propagation algorithms have been developed.[6] It is the best known example of *global constraint*, namely a constraint involving lists of variables with lots of applications and efficient propagation algorithms.)

	1000 S +	100 E +	10 N +	D	+
	1000 M +	100 O +	10 R +	E	=
10000 M +	1000 O +	100 N	10 E +	Y	

The final constraint ensures that the sum of the first two words adds up to the third one. This is enforced by the following arithmetic constraint:

This CSP is satisfiable and admits a unique solution.

3.2.2   Search

Given a CSP over the variables X_1, ..., X_k with domains D_1, ..., D_k, respectively, and a set C of constraints, the goal is to identify its solutions, if any. The *search space* is composed of the sets of all possible combinations of assignments, *i.e.*, $D_1 \times D_2 \times ... \times D_k$.

An algorithm called a *constraint solver* processes the CSP and explores the search space in order to retrieve the solutions. Typically, the search space is unmanageable (*e.g.*, if $|D_i| \geq 2$, then the search space will have size at least 2^k). Therefore, the constraint solver needs to explore the search space efficiently, in particular avoiding the exploration of set of solutions that are incompatible with the given constraints.

A constraint solver is typically composed of two interleaved activities: *constraint propagation* and the solution's actual *search* (often referred to as *labeling*). The first phase represents the *reasoning* step, where the knowledge about the domains and constraints allows the solver to perform some inferences about the problem and, typically, to single out inconsistent sets of solutions. The main idea is repeatedly to consider constraints that may remove values from some variables' domains that surely falsify some of the constraints.

The second phase is a brute force search that is applied when no more propagation is possible. Labeling proceeds by selecting a remaining value from one of the variable's domains, assigning it to the variable and recursively repeating the solving process, starting again with a propagation phase. A comprehensive description of these methods was given by Rossi *et al.*[1]

Example

In the SEND + MORE = MONEY example, the search space consists of 10^8 possible solutions (*i.e.*, the combinatorial combination of eight variables with a domain of 10 elements each). The constraints $S \neq 0$ and $M \neq 0$ immediately allow us to remove one useless value from the domains of S and M. As an example of propagation of the above arithmetic equation, the value of M has to be 1 because it is different from 0 and the carry provided by the sum of S and M cannot be more than 1. Looking at the sum $S + M$, S must be ≥ 8, assuming that $E + O$ can have a carry and $M = 1$.

Other propagations are possible; however let us focus on the two alternative values for S. If $S = 8$, then $O = 0$ (because $S + M = O$ and there is a carry). Since $E + O$ generates a carry, E must be equal to 9, assuming that $N + R$ has a carry. The process continues until a solution is found. Similar reasoning steps can be performed for the case $S = 9$.

3.2.3 Constraint Optimization Problems

A *constraint optimization problem* (COP) is a CSP that includes a cost function to be optimized. The cost function depends on the CSP variables and each solution to the CSP can be evaluated by the cost function. Solving a COP requires the identification of the solution of the CSP which has best cost function value.

For instance, if variables are related to 3D points of a chemical compound, a cost function can score each assignment according to an energetic model encoded in the functional form and associated parameters of a force-field such as CHARMM and AMBER,[7] possibly with an implicit solvent model or of a knowledge-based potential (see, *e.g.*, Lazaridis and Karplus[8] for a general framework and Pokarowski *et al.*[9] for an overview and comparison of most available potentials).

In order to guide the search towards the optimal solution, it is possible to discard CSP solutions that can be proved to be sub-optimal. Several such techniques build on top of the already described constraint propagation and labeling techniques. One of the most commonly used approaches is called *branch and bound*. Given a cost function, the *bound* is a computed estimate of the cost function. The bound is checked when not all variables have been assigned. If the bound is computed to be worse than the current best solution, there is no need to investigate that branch further, since any possible solution will not improve the result.

Branch and bound techniques rely on effective bounds computations. Simple cost functions provide useful bounds that can prune the search space effectively. However, a complex energy landscape of a protein generally provides poor bounds estimations, given the high number of local minima to be explored.

3.3 Constraint Programming-based Applications

The use of constraint programming techniques to address problems in the domain of protein structure prediction commonly requires the introduction of approximated models. The most popular approximations can be divided into three families:

1. *Amino acid representation*: different approximated representations of each amino acid can be devised depending on how many of its atoms are captured by the model.

2. *Space representation*: the three-dimensional space can be approximated
   either using lattice models or using off-lattice models; in the second case, it
   is also possible to distinguish between *a priori* and *a posteriori*
   approximations.
3. *Energy model:* different approximations can be derived based on how the
   free energy of the protein is calculated.

Models representing all of the amino acids' atoms – referred to as *all-atom*
models – are commonly used in molecular dynamics, where small protein
movements are simulated using computational models of physical and
chemical properties. Molecular dynamics is known to be highly computationally intensive, often requiring months of computing time using supercomputers
to represent even relatively simple models. All-atom models are, in general, too
complex to use for simulating large movements of a protein or for *ab-initio*
structure predictions.

At the opposite extreme, the Cα model abstracts an amino acid as the
coordinates of its Cα atom. A distance constraint of 3.81 Å between
consecutive amino acids should be imposed.

Between these two extremes there is a set of intermediate approaches – for
instance, a pair Cα–centroid of the side chain, the full atom list of the
backbone plus a representative of the side chain, and so on.

As far as the space representation is concerned, several crystal lattice models
have been proposed for protein representation.[10] For example, the Cartesian
(or cubic) lattice is characterized by a lattice structure where points are pairs of
integer numbers, connected by edges if and only if their distance is 1. Another
popular lattice structure is the *face-centered cubic lattice* (FCC), which has
been extensively used in constraint-based approaches combined with the Cα
model. Lattice models do not allow knowledge from known protein structures
to be easily re-used, which is crucial for predicting large structures by
assembling smaller, known structures. The main issue is that the accurate
placement of a structure is difficult owing to the rigidity and approximation of
a lattice domain.

Off-lattice models remove the restriction of operating on a structured
representation of the 3D space; on the other hand, they increase the already
huge search space. Even in the case of off-lattice models, some approximations
are needed for representing a 3D point – owing to the limited precision of
information stored in a computer. This can be done *a priori*, *i.e.*, by explicitly
selecting a degree of discretization of the 3D space (*e.g.*, 0.01 Å) or *a posteriori*,
i.e., by choosing a precision in the result of all floating-point computations.
The first solution allows us to make use of finite domain constraint solvers,
whereas the second approach simplifies matrix operations that are typically
employed for dealing with rotations of parts of a protein. We refer to *a priori*
approximated off-lattice models and lattice models as *discrete space models*. *A
posteriori* approximated off-lattice models and other models are instead
referred to as *continuous space models*.

The energy model relies strongly on the amino acid representation chosen. All-atom models have consolidated energy models which have been refined and used extensively in the last decade, whereas for approximated models several potentials have been proposed differing in protein representation, goals (*e.g.*, best separation of native from non-native structures or different flavors of Boltzmann distribution inversion) and by reference state. There are expected correlations among different models, but no consensus (depending on application) has emerged as yet.[9] The simplest model is probably the HP model.[11]

3.3.1 Protein-related Space Models

Let us start by reviewing some basic models used to represent the spatial properties of amino acids. For the sake of simplicity, we represent each amino acid using a single point in space (*e.g.*, using the Cα model). Let us consider the following sequence of amino acids: acdef. Let us define five corresponding variables (*A, C, D, E, F*) and show different domains and constraints to model the peptide main geometric features.

3.3.1.1 Continuous Domains

The five variables are associated with domains that range over a portion of the continuous space: $D_A = D_C = D_D = D_E = D_F = [(0, 0, 0), (100, 100, 100)]$. This means that each point can range in a box of side 100. See also Figure 3.2.

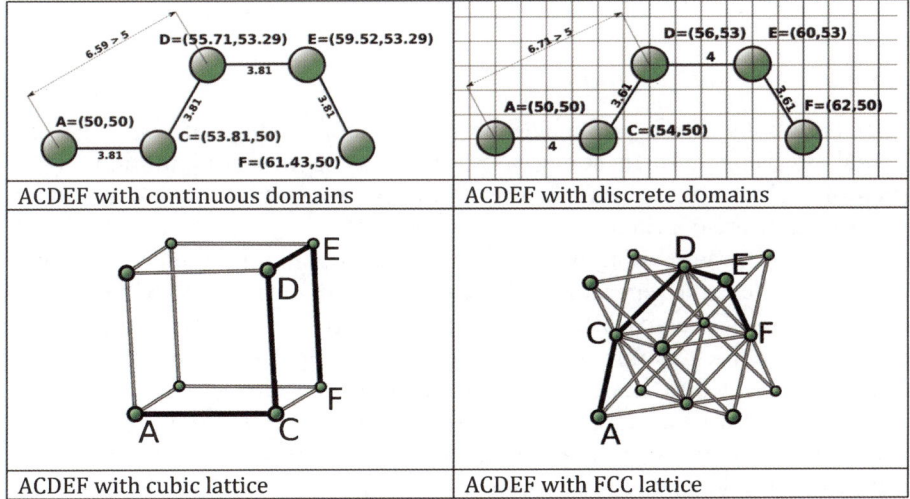

ACDEF with continuous domains	ACDEF with discrete domains
ACDEF with cubic lattice	ACDEF with FCC lattice

Figure 3.2 Comparison of domain representations.

The first constraint models the distance between two consecutive Cα atoms ($\sim$3.81 Å). The relational constraint

$$dist(P,\ Q,\ r,\ R)$$

represents a *distance* constraint between the two atoms. This constraint will discard all the positions of P and Q that violate the following relation:

$$r \le \sqrt{(P_x - Q_x)^2 + (P_y - Q_y)^2 + (P_z - Q_z)^2} \le R$$

where P_x, P_y and P_z are the (x,y,z) coordinates of P.

In order to allow for a level of tolerance in the placement of the amino acids, we typically impose the following constraint:

$$dist(A,\ C,\ 3.79,\ 3.83),\ dist(C,\ D,\ 3.79,\ 3.83),$$

$$dist(D,\ E,\ 3.79,\ 3.83),\ dist(E,\ F,\ 3.79,\ 3.83)$$

Moreover, a steric constraint should be enforced to avoid the collapse of the peptide. Each pair of non-consecutive amino acids can be constrained to be at a minimum distance. For example, the constraint $dist(A,\ E,\ 5,\ 100)$ states that A and E are at a minimum distance of 5 Å, while the maximum distance is a large overestimation. A common practice is to eliminate some symmetries by fixing the positions of the first amino acids, *e.g.*, $A = (50, 50, 50)$ and $C = (53.81, 50, 50)$. Such an initial model can be refined with the addition of constraints about pseudo-torsional angles, side chain mobility, secondary structure information, *etc.*

3.3.1.2   Discrete Domains

The continuous model presented above has the advantage of using the best spatial precision and, in some cases, can lead to fast computation of the solutions. However, the presence of non-linear constraints makes the search non-trivial in the continuous case.

Constraint programming benefits from reasoning about a finite and discrete search space (even if very large). This capability can be achieved by an *a priori* discretization of the space that is compatible with the quality of the desired resolution. In the previous example, one could adopt a quantization of 0.01 Å to produce a discrete version of the 3D space.

For comparative purposes, let us consider the extreme case of a discretization of the search space based on a resolution of 1 Å. This would lead, for example, to the following constraints:

$$D_A = D_C = D_D = D_E = D_F = (0, 0, 0) \ldots (100, 100, 100)$$

$$dist(A, C, 3, 5), dist(C, D, 3, 5)$$

$$dist(D, E, 3, 5), dist(E, F, 3, 5)$$

$$A = (50, 50, 50), C = (54, 50, 50)$$

$$dist(A, D, 5, 100), dist(A, E, 5, 100), dist(A, F, 5, 100)$$

$$dist(C, E, 5, 100), dist(C, F, 5, 100), dist(D, F, 5, 100)$$

In Figure 3.2 (top right), we show a possible solution to these constraints. Note that the coarse level of discretization requires the use of looser bounds on the distances, leading to more approximated solutions. However, the techniques developed for finite domain propagation can lead to the solutions in a shorter time.

3.3.1.3 Lattice Domains

At the end of the spectrum of discretization, *crystal lattice models* can be used to represent a discrete and regular partition of the space in a convenient way.

In the literature there is a large collection of crystal lattice spatial models.[12–14] A lattice has regular properties and can be modeled by *lattice constraints*. Each 3D point is mapped to one of the points in the lattice.

The domains of the amino acids can be described by a simple relation that depends on the shape of the lattice. For a cubic lattice, the relation is

$$P \in \{(x,y,z), x \in Z, y \in Z, z \in Z\}$$

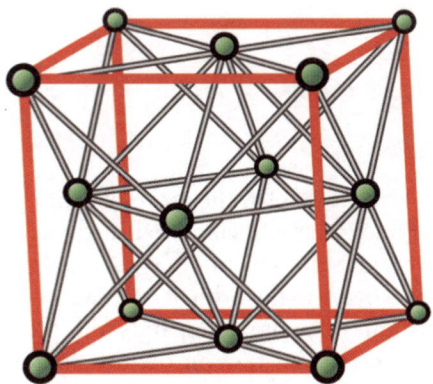

Figure 3.3   A cubic lattice (solid red lines) and FCC lattice (thin gray lines).

and an FCC lattice can be described by the following set of points:

$$P \in \{(x,y,z),\ x+y+z \text{ is even},\ x \in Z,\ y \in Z,\ z \in Z\}$$

An example of the periodic unit of cubic and FCC lattices is shown in Figure 3.3. The solid red lines connect the vertexes of the cubic unit (for comparison purposes, the side has length equal to 2 in order to match the FCC unit). The thin gray lines connect the vertexes that have an even sum of the coordinate components. Depending on the representation of the amino acids, one or several points can be used to represent each amino acid and these have to be mapped to the lattice.

A convenient way to model consecutive amino acids is to constrain the distance in terms of *lattice units*. Two consecutive amino acid points P and Q can be separated by one lattice unit (*next constraint*). For the cubic lattice, the next constraint is

$$|P_x - Q_x| + |P_y - Q_y| + |P_z - Q_z| = 1$$

and for the FCC lattice it is

$$|P_x - Q_x| + |P_y - Q_y| + |P_z - Q_z| = 2$$

The lattice unit is scaled in order to match the actual distance between two points. When modeling a backbone made of Cαs only, the typical distance of 3.81 Å is scaled to 1 in the case of a cubic lattice and to $\sqrt{2}$ in the case of an FCC lattice.

Some *global* constraints can be defined in terms of properties of the whole protein. Lattice domains allow global constraints to be handled in a convenient way. For instance, when modeling backbone properties, the *alldistant* and *self-avoiding walk* constraints can be modeled in terms of alldifferent and lattice unit distance constraints. Dal Palù et al.[15] reported and formalized some examples of these.

3.3.2 Lattice Space Models

3.3.2.1 HP Model

Let us consider the Cα abstraction for the representation of amino acids. In the HP energy model,[11] amino acids are classified in two categories: *hydrophobic* (H) and *polar* (P). Two hydrophobic amino acids in contact provide an energy contributions of –1, whereas the other contacts do not provide any contributions. The notion of contact depends on the space representation adopted. It is common practice to rely on defining a contact as a distance of one or two lattice units, which corresponds to a distance of roughly 5 Å. Observe that, even in presence of relatively simple lattices (*e.g.*, the Cartesian lattice, where a distance of 1 is chosen as the *contact* distance), the problem of finding a folding with a global energy smaller than a given value k is *NP-complete* (a class of hard problems to be solved), as proved by Crescenzi *et al.*[16]

Backofen and Will investigated this problem using the more realistic FCC lattice and solved it successfully using constraint programming techniques for sequences of length 160 and more.[17,18] Efficiency is obtained using clever symmetry breaking techniques to reduce the search space and by introducing the notion of *core*. If we consider the protein in a 3D box, the folding is analyzed layer by layer and the conformations in each layer that maximize the number of contacts are pre-computed. This kind of approach cannot be immediately extended to more detailed energy models. Moreover, if additional structural constraints, *e.g.*, coming from secondary structure information, are introduced, then the core pre-computation will become inappropriate. Nevertheless, this is one of the most successful *ab initio* approaches for lattice model prediction, at least from the computer science perspective – where a huge search space is completely visited and the best solution is found. Several other researchers faced the same problem, using approximated techniques. In particular, Shmygelska and Hoos[19] provided approximated solutions to the same problem using local search and meta-heuristics techniques. Cebrian *et al.*[20] and Dotú *et al.*[21] proposed approaches to mix local search and constraint programming. The positive aspect of these contributions is that they can be generalized to different energy functions and lattice models; their drawback, however, is that they cannot find the optimum already found by Backofen and Will, even using large computing resources and time.

Let us conclude this section by noting that slightly more complex energy models have been proposed by Backofen and Will for the protein structure prediction problem in the FCC lattice. They considered an energy model in which amino acids are classified in four categories (HPNX).[22] However, this preliminary proposal has not been further developed.

3.3.2.2 Protein Docking

Barahona and Kripphal[2,23] provided a constraint-based solution to the problem of predicting the interesting relationships between two or more protein surfaces. This problem is an instance of the *protein docking problem*. In this case, both energetic and spatial interactions guide the selection of the functional and actual pose. The authors use constraint programming for selecting compatible matchings between the rigid surface shapes of different proteins. The matchings are scored using a precise force-field analysis. This approach relies on the use of an off-lattice space model that is discretized *a priori* into small cubes. The resulting tool, called *Chemera*, produces results comparable to other approaches, but with significantly less computational effort (in terms of both time and memory requirements).

3.3.2.3   Phase Reconstruction

Heldt and Bockmayr[24] addressed the phase reconstruction problem using data derived from X-ray crystallographic analysis. The goal is to generate an electron density map that describes the protein shape. The problem is that X-ray diffraction experiments provide only the intensities of the X-rays diffracted in different directions, and the information about the phase shift of the corresponding waves is lost. The authors showed how this information can be retrieved by elegantly solving a difficult constraint satisfaction problem. The approach relies on the use of a discrete grid (*i.e.*, a Cartesian lattice), where a density map is represented by density values at grid points. They tackled the problem by using binary envelopes of the molecule, *i.e.*, a binary function on the lattice, assigning value 1 to lattice points where the electron density is above a selected threshold. This allows the phase problem to be cast as a collection of linear binary constraints.

3.3.2.4   COnstraint Programming on LAttices (COLA)

COnstraint programming on LAttices (COLA)[25] is a general framework for constraint solving on FCC lattices that has been used to investigate the protein structure prediction problem, using a Cα amino acid model. The difference with respect to the approach proposed by Backofen and Will[17,18] is in the energy model; COLA makes use of a 20×20 statistical potential contact energy model, developed by Berrera *et al.*[26] Two amino acids are considered in *contact* if their distance is less than a selected threshold. In the COLA FCC model, each lattice unit has length $\sqrt{2}$, which for convenience is scaled to correspond to 3.8 Å; thus, a contact distance of 2 in the lattice scale guarantees that amino acids are in contact if they are separated by at most 5.4 Å.

COLA allows also the user to suggest secondary structure information – note that the contact energy functions on FCC lattice models are, in general, unable to render the 3.6 amino acids per turn required to model helices in real proteins.

COLA represents an evolution of previous approaches modeling the protein structure prediction problem using constraint solving on FCC lattices – predominantly based on the use of existing constraint solvers (*e.g.*, constraint logic programming).[27,28] In particular, COLA provides an *ad hoc* constraint solver, with the following main features:

- The domain of a variable is a set of lattice points, represented as a 3D interval (*i.e.*, a box).
- Some global constraints are implemented as built-in, such as *alldistant* and *rigid block* – these allow us, for example, to pre-define known sub-parts of the protein (*e.g.*, secondary structure components).
- The solver supports search heuristics justified by the protein structure prediction problem.

In particular, the *bounded block fail* (BBF) heuristic proved to be effective for the protein structure prediction problem. Intuitively, BBF partitions the amino acids in blocks and limits the number of consecutive search steps that can be performed within a block. The key idea is that most small local changes do not significantly change the form of a protein. When a sufficient number of attempts have been made looking at very close conformations without success, it is reasonable to abandon that research branch.

COLA has demonstrated good performance in predicting good-quality approximations of protein structure (with minimal or almost minimal energy) on FCC lattices, *e.g.*, Dal Palù *et al.*[25] reported experiments with proteins of length up to 100 amino acids, computed within a reasonable computation time. COLA is available on-line from www.cs.nmsu.edu/fiasco.

COLA has been adopted (and extended) by other researchers; *e.g.*, the methodology described by Ullah and Steinhöfel[29] builds on COLA to generate the neighborhood conformations that are to be used in generic local search procedures for protein conformation discovery.

3.3.2.5 *Portfolios for the Prediction Problem*

Arbelaez *et al.*[30] explored the use of machine learning algorithms to select the heuristics for visiting the conformational space of a protein. In particular, they tested and analyzed the impact of different search heuristics offered by the constraint programming platform *Gecode* (www.gecode.org).

The investigation relies on the use of a Cartesian model for the representation of the space and on the use of the Cα model for the representation of amino acids. The energy model is based on the classification of amino acids in three groups according to four properties (hydrophobicity, volume, polarity and polarizability). The study is based on performing independent computations with a short time-out and evaluating the performance depending on the chosen parameters. The most promising strategy is then used to perform a complete search.

Although the work of Arbelaez *et al.*[30] is based on a discrete spatial model, the same technique can also be used on continuous models.

3.3.3 Off-lattice Space Models

3.3.3.1 *NMR Constraints*

Krippahl and Barahona focused on modeling protein structure as a combination of geometric constraints and other constraints coming from experimental data (*e.g.*, NMR data).[2,31] Common protein structure prediction algorithms for simulated annealing and torsion angle dynamics[32] hide the set of soft and hard constraints in the energetic terms. In contrast, with constraint programming such constraints are made explicit in the model of the geometry of the system.

The system proposed by Barahona and Krippahl, called *PSICO* (Processing Structural Information with Constraint programming and Optimization), encodes constraints derived from X-ray crystallography and NMR spectroscopy. As such, PSICO cannot be considered an *ab initio* predictor. The typical forms of such constraints are

- distance constraints between pairs of atoms and
- angular constraints about the relative orientation of inter-atomic bonds.

As in COLA, the authors model a point as a triple of coordinates, but in this case using a continuous space model. The domain of each point is a 3D cuboid, which allows the solver to *propagate* inter-atomic distance constraints and to introduce some global constraints concerning rigid groups of atoms.

3.3.3.2 Fragment Assembly with Constraints

The main limitation of a lattice-based approach is the difficulty of integrating and/or reusing information from repositories of known protein structures. The success of systems such as *Rosetta*[33] has demonstrated the importance of being able to integrate these types of information to achieve fast and effective structure prediction. Local information carried by homologous fragments can be exploited by a tool that combines potential candidates into a putative protein to be evaluated from an energetic point of view. Segments of the protein are associated with a set of possible 3D fragments; the prediction problem can then be reduced to the problem of assembling these fragments into feasible structures.

State-of-the-art prediction tools such as Rosetta rely on local search techniques to support the assembly stage. More recently, Dal Palù *et al.*[34] presented a tool, constructed using constraint logic programming, for fragment assembly. In particular, a selection of non-redundant peptides from the Protein Data Bank is used to create a database of known conformations for all sequences of four amino acids. The problem of fragment assembly is described as a CSP based on a discrete spatial model with resolution of 0.1 Å. Each amino acid is modeled by two centers: the Cα atom and the center of mass of the side chain (*centroid*). Distance constraints are based on each amino acid's side chain occupancy.

Putative solutions correspond to assemblies of fragments that respect the distance constraints and do not create any clashing. Non-clashing solutions are scored using an energy function, in order to locate those conformations that provide a minimal global energy. The energy function adopted is characterized by three components: a statistical contact potential for side-chain and backbone interactions, a pseudo-torsional angle component that accounts for statistically observed preferences in the structure of the fragments and a component that considers the relative orientations of spatially close triplets of amino acids. Several heuristics have been developed to guide the search for components to be assembled; however, an *ad hoc* implementation of a large neighboring search (also known as constraint-based local search) proved to be the most effective approach.[35]

The approach was used effectively in the study of the estrogen receptor behavior in the case of large time-scale flexibility and interaction with ligands.[36] More specifically, constraints that model the receptor are imposed in order to describe the rigid parts of the protein as deposited in the PDB. Additional constraints are added in order to model a flexible loop and its mobility (in terms of spatial domain). An exhaustive search allows millions of different poses to be retrieved. In this case, only geometric constraints are enforced and no energetic functions are used to evaluate the protein–ligand complex. However, some further analysis allows us to track the most likely geometric dynamics that connect two specific end-points and that pass through the poses found as a solution of the CSP.

The project on constraint-based fragment assembly has developed from its initial constraint logic programming implementation, presented by Dal Palù *et al.*,[34] into a comprehensive tool, referred to as *Fragment Interactive Assembly for protein Structure with COnstraints (FIASCO)*.[37] FIASCO explored several new directions, ranging from the development of new approaches for modeling the protein structure, to new constraint-solving solutions, to the creation of interaction models that allow the participation of the scientist in the structure determination process.

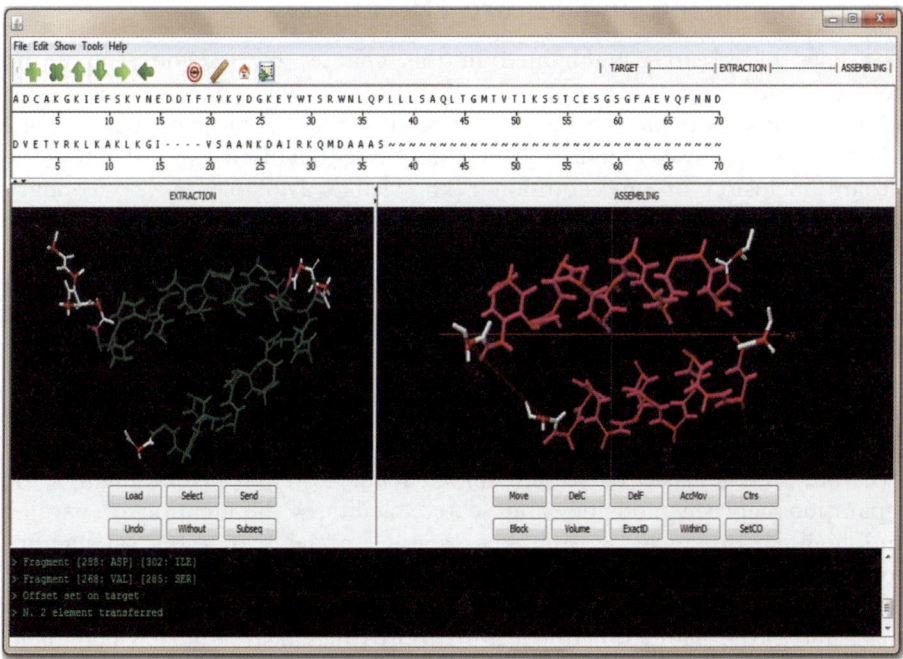

Figure 3.4 A screenshot of FIASCO.

As far as the interface is concerned, it provides the user with several operative options. FIASCO supports the traditional *ab initio* determination of a protein structure with minimal global energy for a given sequence of amino acids. In addition, FIASCO allows the user to search protein databases for long fragments of the protein that have a known structure, and also to perform secondary structure prediction. These fragments and secondary structure components appear in the left window of FIASCO's user interface (Figure 3.4). The user, using the mouse, can decide the relative positions of these fragments. The assembling of all the fragments is performed by the constraint solver, using techniques analogous to those mentioned earlier and including all the information provided (including relative spatial positions of fragments). Each 3D structure computed by the constraint solver is presented to the users, allowing them to modify it, *e.g.*, by moving some of the parts and restarting the solving process.

As far as the modeling part is concerned, the former Cα–centroid model has been extended with a model that accounts for all the atoms of the backbone (including the Cα itself) plus the centroid of the side chain. This allows us to use a more precise energy function, specifically designed for that work. The constraint solver itself has been implemented using optimized matrix transformation operations and supports the parallel exploration of the search space on Beowulf clusters.

3.4   Conclusion and Future Directions

The lines of research highlighted in this chapter have demonstrated the potential advantages of applying constraint programming techniques to address problems related to the prediction and analysis of protein structures. The separation between modeling and solution search offered by CP allows the elegant inclusion of any additional knowledge available (*e.g.*, secondary structure components and/or homologies) in guiding the prediction of the structure of the complex. It also permits an incremental refinement of the solutions, allowing a '*scientist-in-the-loop*' approach – where the scientist is able to interact with the prediction process, by reviewing partial solutions and dynamically introducing preferences and additional knowledge.

The process used to explore the search space can itself be explicitly programmed and thus customized to the needs of each specific protein being studied. Different heuristics can be used and dynamically selected based on the structure of the specific search space being explored. Additionally, the separation between modeling and search facilitates the creation of parallel and high-performance algorithms to speed up the search for satisfactory conformations.

The use of constraints and simplified representations of amino acids, 3D space and of the energy function lead to approximated structures, which might need to be further refined using more traditional techniques (*e.g.*, molecular dynamics simulations). Nevertheless, the use of CP allows the exploration of a large search space and the exploration is not bound by a temporal evolution process as in the case with dynamic simulations.

Although the investigation of CP approaches to protein structure prediction was originally demonstrated on relatively small proteins, there has recently been a greater emphasis on *scalability*, in order to permit the application of the proposed techniques to the investigation of actual proteins with unknown structures. Several studies have recently emerged in this direction, *e.g.*, Best *et al.*[37] explored the use of CP techniques to study proteins in the inner ear of the *Xenopus laevis*.

The application of constraint technology to protein modeling and to more general bioinformatics problems has also been promoted by an international workshop series that we co-organized during the last 8 years: WCB (Workshop on Constraint-based Methods for Bioinformatics; www.bioinf.uni-freiburg.de/Events/WCB12/), and we refer the interested reader to the online proceedings of the various editions.

Acknowledgements

We are grateful for grants MIUR FIRB RBNE03B8KK, GNCS-INdAM 2009-10-11, PRIN 2007M3E2T2, PRIN 20089M932N, NIH P50-GM68762 and NSF IIS-0812267, CBET-0754525 and to the US Army High Performance Computing Research Center.

References

1. F. Rossi, P. Van Beek and T. Walsh, *Handbook of Constraint Programming*, Elsevier, Amsterdam, 2006.
2. L. Krippahl and P. Barahona, PSICO: solving protein structures with constraint programming and optimisation, *Constraints*, 2002, **7**, 317–331.
3. K. Marriott and P. Stuckey, *Programming with Constraints: an Introduction*, MIT Press, Cambridge, MA, 1998.
4. J. Jaffar and M. Maher, Constraint logic programming: a survey. *J. Logic Programming*, 1994, **19–20**, 503–581.
5. P. Van Hentenryck and L. Michel, *Constraint-based Local Search*, MIT Press, Cambridge, MA, 2005.
6. J.-C. Règin, A filtering algorithm for constraints of difference in CSPs, in *Proceedings of the 12th National Conference on Artificial Intelligence (AAAI)*, 1994, pp. 362–367.
7. A. D. Mackerell, Empirical force fields for biological macromolecules: overview and issues, *J. Comput. Chem.*, 2004, **25**, 1584–1604.
8. T. Lazaridis and M. Karplus, Effective energy functions for protein structure prediction, *Curr. Opin. Struct. Biol.*, 2000, **10**(2), 139–145.
9. P. Pokarowski, A. Kloczkowski, R. L. Jernigan, N. S. Kothari, M. Pokarowska and A. Kolinski, Inferring ideal amino acid interaction forms from statistical protein contact potentials, *Proteins*, 2005, **59**(1), 49–57.

10. J. Skolnick and A. Kolinski, Reduced models of proteins and their applications, *Polymer*, 2004, **45**, 511–524.
11. K. A. Dill, Dominant forces in protein folding, *Biochemistry*, 1990, **29**, 7133–7155.
12. A. Kolinski, P. Rotkiewicz, B. Ilkowski and J. Skolnick, Protein folding: flexible lattice models, *Prog. Theor. Phys.*, 2000, **138**, 292–300.
13. L. Toma and S. Toma, Folding simulation of protein models on the structure-based cubo-octahedral lattice with the contact interactions algorithm, *Protein Sci.*, 1999, **8**(1), 196–202.
14. A. Dal Palù, Constraint programming approaches to the protein structure prediction problem, *PhD Dissertation*, University of Udine, 2006.
15. A. Dal Palù, A. Dovier and E. Pontelli, Computing approximate solutions of the protein structure determination problem using global constraints on discrete crystal lattices, *Int. J. Data Mining Bioinformatics*, 2010, **4**(1), 1–20.
16. P. Crescenzi, D. Goldman, C. Papadimitriou, A. Piccolboni and M. Yannakakis, On the complexity of protein folding, *J. Comput. Biol.*, 1998, **5**(3), 423–466.
17. R. Backofen and S. Will, Excluding symmetries in constraint-based search, *Constraints*, 2002, **7**(3–4), 333–349.
18. R. Backofen and S. Will, A constraint-based approach to fast and exact structure prediction in 3-dimensional protein models, *Constraints*, 2006, **11**(1), 5–30.
19. A. Shmygelska and H. H. Hoos, An ant colony optimisation algorithm for the 2D and 3D hydrophobic polar protein folding problem, *BMC Bioinformatics*, 2005, **6**, 30.
20. M. Cebrian, I. Dotu, P. Van Hentenryck and P. Clote, Protein structure prediction on the face centered cubic lattice by local search, in *Proceedings of the National Conference on Artificial Intelligence (AAAI)*, 2008, pp. 241–245.
21. I. Dotú, M. Cebrián, P. Van Hentenryck and P. Clote, Protein structure prediction with large neighborhood constraint programming search, in *Proceedings of the International Conference on Principles and Practice of Constraint Programming (CP)*, LNCS 5202, Springer, Berlin, 2008, pp. 82–96.
22. R. Backofen, S. Will and E. Bornberg-Bauer, Application of constraint programming techniques for structure prediction of lattice proteins with extended alphabets, *Bioinformatics*, 1999, **15**(3), 234–242.
23. P. Barahona and L. Krippahl, Constraint programming in structural bioinformatics, *Constraints*, 2008, **13**, 3–20.
24. C. Heldt and A. Bockmayr, Geometric constraints for the phase problem in X-ray crystallography, un *Proceedings of the Workshop on Constraint Based Methods in Bioinformatics*, 2010, pp. 36–41.
25. A. Dal Palù, A. Dovier and E. Pontelli, A constraint solver for discrete lattices, its parallelization and application to protein structure prediction, *Software Pract. Experience*, 2007, **37**(13), 1405–1449.

26. M. Berrera, H. Molinari and F. Fogolari, Amino acid empirical contact energy definitions for fold recognition in the space of contact maps, *BMC Bioinformatics*, 2003, **4**, 8.
27. A. Dal Palù, A. Dovier and F. Fogolari, Constraint logic programming approach to protein structure prediction, *BMC Bioinformatics*, 2004, **5**, 186.
28. A. Dal Palù, A. Dovier and E. Pontelli, Heuristics, optimizations and parallelism for protein structure prediction in CLP(FD), in *Proceedings of the Conference on Principles and Practice of Declarative Programming (PPDP)*, ACM Press, New York, 2005, pp. 230–241.
29. A. Ullah and K. Steinhöfel, A hybrid approach to protein folding problem integrating constraint programming with local search, *BMC Bioinformatics*, 2010, **11**(Suppl. 1), S39.
30. A. Arbelaez, Y. Hamadi and M. Sebag, Building portfolios for the protein structure prediction problem, presented at the Workshop on Constraint Based Methods for Bioinformatics (WCB), Edinburgh, July 2010.
31. L. Krippahl and P. Barahona, Propagating n-ary rigid-body constraints, in *Proceedings of the International Conference on Principles and Practice of Constraint Programming (CP)*, LNCS 2833, Springer, Berlin, 2003, pp. 452–465.
32. P. Güntert, C. Mumenthaler and K. Wüthrich, Torsion angle dynamics for NMR structure calculation with the new program DYANA, *J. Mol. Biol.*, 1997, **273**, 283–298.
33. S. Raman, R. Vernon, J. Thompson, M. Tyka, R. Sadreyev, J. Pei, D. Kim, E. Kellogg, F. DiMaio, O. Lange, L. Kinch, W. Sheffler, B.-H. Kim, R. Das, N. V. Grishin and D. Baker, Structure prediction for casp8 with all-atom refinement using Rosetta, *Proteins*, 2009, **77**(S9), 89–99.
34. A. Dal Palù, A. Dovier, F. Fogolari and E. Pontelli, CLP-based protein fragment assembly, *Theory Pract. Logic Programming*, 2010, **10**(4–6), 709–724.
35. N. Mladenovic and P. Hansen, Variable neighborhood search, *Comput. Operations Res.*, 1997, **24**, 1097–1100.
36. A. Dal Palù, F. Spyrakis and P. Cozzini, A new constraint logic programming-based approach for investigating the protein flexibility. A first application to the estrogen receptor case, *Eur. J. Med. Chem.*, 2012, **49**, 127–140.
37. M. Best, K. Bhattarai, F. Campeotto, A. Dal Palù, H. Dang, A. Dovier, F. Fioretto, F. Fogolari, T. Le and E. Pontelli, Introducing FIASCO: fragment-based interactive assembly for protein structure prediction with constraints, presented at the Workshop on Constraint-Based Methods in Bioinformatics (WCB), Perugia, September 2011.

CHAPTER 4

Molecular Dynamics: a Tool to Understand Nuclear Receptors

FRANCESCA SPYRAKIS*[a,b], XAVIER BARRIL*[c,d] AND
F. JAVIER LUQUE*[c]

[a] Dipartimento di Chimica Generale ed Inorganica, Chimica Analitica,
Chimica Fisica, Università degli Studi di Parma, Parma, Italy; [b] INBB,
Biostructures and Biosystems National Institute, Rome, Italy; [c] Departament
de Fisicoquímica and Institut de Biomedicina (IBUB), Facultat de Farmàcia,
Universitat de Barcelona, Barcelona, Spain; [d] Institució Catalana de Recerca i
Estudis Avançats (ICREA), Barcelona, Spain
*E-mail: francesca.spyrakis@unipr.it; xbarril@ub.edu or fjluque@ub.edu

4.1 Nuclear Receptors: Overview of the Ligand Binding Domain

The biological activities encompassed by the members of the nuclear receptor (NR) superfamily are very rich and diverse, as they comprise the formation of interactions with other biological components, rearrangements of the quaternary structure, dimerization, response to post-transcriptional modification and binding or release of signaling molecules, amongst others, as noted in Chapter 1. All these processes are highly dynamic and involve a wide range of structural adaptations. Therefore, understanding the function of nuclear receptors at the molecular level necessarily requires not only the structural characterization of NR components, but crucially also the determination of their ability to react to different stimuli. Whereas experimental techniques are almost exclusively responsible for the elucidation of the structural features of

RSC Drug Discovery Series No. 30
Computational Approaches to Nuclear Receptors
Edited by Pietro Cozzini and Glen E. Kellogg
© The Royal Society of Chemistry 2012
Published by the Royal Society of Chemistry, www.rsc.org

NRs, computational techniques and particularly molecular dynamics (MD), are an essential component for establishing the linkage between structure and function.

X-ray crystallography has played, and continues to play, a fundamental role in the structural characterization of NRs. However, a major limitation of this technique is that the molecules under study must be able to adopt sufficiently compact and rigid structures to pack and form crystals. Accordingly, large numbers of structures have been solved for the ligand-binding domain (LBD) and the DNA-binding domain (DBD), which are the most rigid segments. By contrast, it has not been possible to obtain structures of the N-terminal domains, which contain the ligand-independent activation function (AF-1), or the hinge region connecting the DBD and LBD. In contrast to LBD and DBD, these domains are much more variable and their structure is not well defined, but they confer a necessary level of structural plasticity. Efforts to characterize the full-length structures, and also biologically relevant multimeric complexes, are under way and they involve the combination of multiple experimental and computational techniques.[1]

Even for those domains with a well-defined structure, namely both DBD and LBD, structural transitions are still required to accomplish the biological functions exerted by NRs. This is particularly well exemplified in the case of the LBD. Accordingly, we focus on this key regulatory element to illustrate the role of dynamics and the suitability of MD simulations to deepen our understanding of this fundamental property.

The LBD (illustrated by PDB entry 1T7T[2] in Figure 4.1) is the site of hormone binding, co-regulatory protein interactions (*i.e.*, the AF-2 surface) and is a target for small-molecule antagonists. As such, it is of great

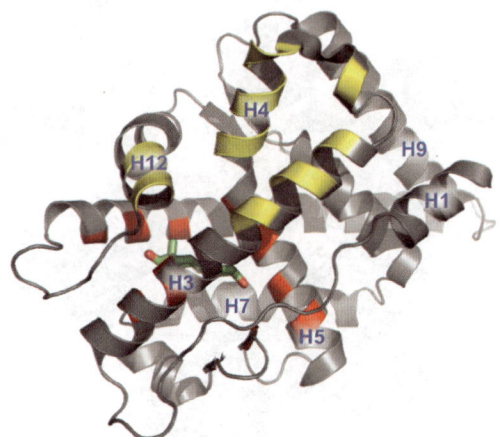

Figure 4.1   Representative structure of the ligand-binding domain of nuclear receptors. The ligand binding site and the AF-2 are highlighted in red and yellow, respectively. Some secondary structures elements are labeled.

pharmacological interest and hundreds of crystal structures exist for the LBD bound to both agonists and antagonists, and also to co-regulatory peptides. The domain generally consists of 11–13 α-helices, depending on the specific receptor. They are folded into a three-layer helical sandwich, which creates a hydrophobic pocket where the ligands bind. The central location of the binding site allows the ligands to interact simultaneously with multiple structural elements, including helices H3, H5, H6, H7, H10 and H12. On the outer surface, several interaction sites exist, of which the AF-2 region, formed by helices H3, H4 and H12, is particularly important because it binds various co-regulators.

Ligand binding implies a major change for the receptor and the apo form of the protein is often difficult to crystallize owing to inherent flexibility. As the binding site is mostly hydrophobic, the residues lining it are likely to collapse, while the secondary structures may become unstructured. This is exemplified in Figure 4.2, where the apo and holo forms of REV-ERBβ are superimposed (PDB entries 3CQV[3] and 2V0V[4]). The flexibility of the binding site also explains the ability to recognize ligands of different shapes and sizes and even the same ligand in different binding modes.[5,6] In fact, the structural plasticity that allows such promiscuity of the binding sites is likely a result of evolutionary adaptation, as NRs often bind several ligands, albeit with different affinities.

In an analogous manner, the structural elements forming the AF-2 site also undergo major conformational changes to accommodate a variety of co-activators and co-repressors. Owing to its close proximity to the ligand-binding

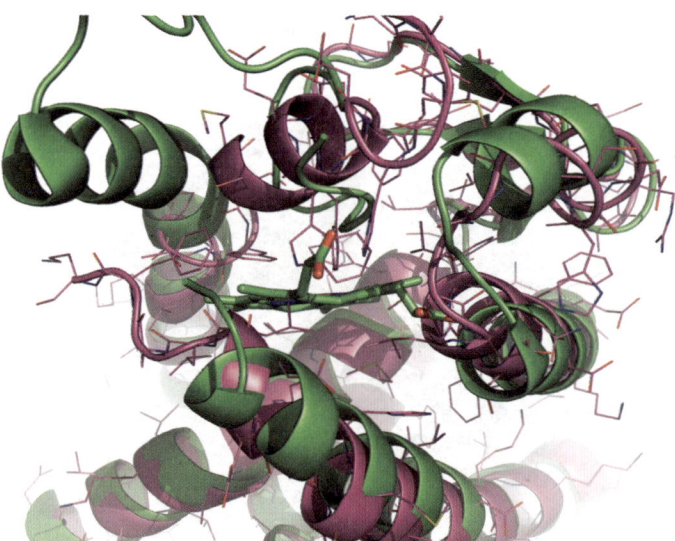

Figure 4.2   Structure of REV-ERBβ in complex with a heme group (holo form; green) and in the apo form (purple). Side-chain and backbone conformational changes allow a change of pocket size of ∼500 Å³.

site, the structural preferences of the AF-2 and, therefore, the relative affinities for different co-regulators are intimately linked to the nature of the ligand. As an example, we quote here a recent study by Le Maire *et al.* focused on the structural basis of ligand-dependent and ligand-independent co-repressor recruitment by human retinoic acid receptor (RAR).[7] The authors found that a short C-terminal region of helix H10 adopts a β-strand conformation (S3) that specifically interacts with CoRNR1 β1 residues, as shown in Figure 4.3, which displays the structures of RAR complexes with (i) an agonist and the binding motif of a co-activator, (ii) an inverse agonist and the binding motif of a co-repressor and (iii) an antagonist that prevents recruiting any co-regulator (PDB entries 3KMR, 3KMZ and 1DKF, respectively).[7,8] The formation of this β-sheet interface is important for increasing the availability of the co-regulator groove of apo RAR by removing helix H12. Furthermore, a novel secondary-structure transition from β-strand S3 to β-helix H11 is the master regulator of co-repressor dissociation from RAR, whereas H12 is primarily involved in the interaction with co-activators.

These are just a few examples demonstrating how critical LBD flexibility is for the function of the NR. Crystal structures have been essential to establish structure–function relationships. However, these static pictures still provide an incomplete view of the system, which, in fact, has access to a whole range of conformations. The biological function is, therefore, not attributable to a single conformation, but to a particular composition of the ensemble, which can be shifted towards one state or another, but is always heterogeneous and in dynamic equilibrium.

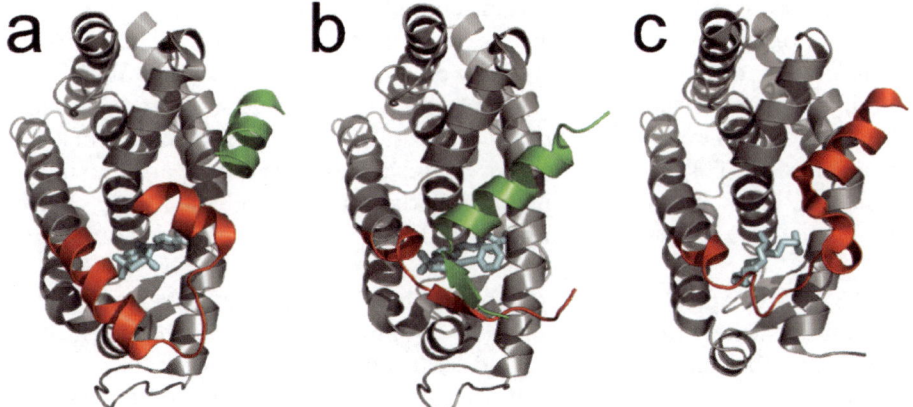

Figure 4.3 Structures of RAR LBD in complex with agonist, inverse agonist or neutral antagonist ligands. (a) Structure of the RAR LBD–SRC-1 NR2–Am580 complex. (b) Structure of the RAR LBD–N-CoRNR1–BMS493 complex. (c) Structure of the RAR LBD–BMS614 complex. The flexible part of the protein (end of H11 and H12) is shown in red, the co-regulator peptide [co-activator for (a); co-repressor for (b)] in green and the ligand in cyan.

Our aim here is to illustrate the impact of MD-based techniques in providing clues that permit the structural features of NRs and their ligands to be related to the biological activity, through conformational alterations triggered upon complex formation. To achieve this goal, we start with a brief description of selected computational tools valuable for gaining insight into the molecular features of ligand recognition and binding. Then, a series of representative studies are discussed in order to analyze the information gained from computational studies regarding three main aspects: ligand specificity, ligand unbinding and allosterism.

4.2 Molecular Dynamics and Protein Flexibility

Classical MD is a powerful strategy to explore the inherent flexibility of proteins.[9–11] Using experimentally determined conformations as the starting point, MD simulations have the ability to generate a whole ensemble of conformations that is a faithful representation of the behavior of the protein in solution.[12,13] Sampling of the conformational space of a molecular system is achieved by iterative integration of Newton's equation of motion, yielding a trajectory that compiles the time evolution of accessible conformational states under specific simulation conditions. To this end, the position and velocity of the particles present in the system are iteratively solved by numerical calculation of instantaneous forces, which can be evaluated in response to the interactions quantified by the force field. MD can be routinely applied to investigate a wide range of dynamic processes, although applications are largely limited in practice by the time scale of the dynamic processes, which in biomolecular systems can range from femtoseconds to hours.

Since atomistic MD simulations are well suited to study elastic motions of atoms/groups of atoms and rotations of side chains, they are valuable in identifying small structural rearrangements triggered upon ligand binding. This can be achieved with current MD practices because such events are usually fast (<100 ns). However, recent advances in the field are making it possible to carry out simulations beyond the microsecond time scale, which allows the investigation of larger amplitude and slower conformational changes.[14–18]

MD also offers a valuable complement to high-resolution experimental techniques, such as time-resolved X-ray crystallography, which has been used to explore the migration of ligands through inner cavities,[19] or NMR techniques, which can provide information about ensembles of low-energy conformations.[20] The synergistic complementarity between these techniques has proven to be extremely successful, as noted in the implementation of NMR-derived data for exploring folding pathways of proteins,[21] to characterize the conformational states corresponding to natively unfolded proteins,[22] and the conformational equilibrium between major substates of RNase A.[23]

4.2.1 Prediction of Binding Affinities by Free Energy Calculations

A qualitative understanding of the chemical features that contribute to drug binding is valuable for the analysis of structure–activity relationships. However, a quantitatively accurate estimate of the binding affinity is required in lead optimization. The use of free energy calculations in conjunction with classical simulations is one of the most powerful approaches for predicting changes in binding affinities, arising due to small chemical differences between structurally related compounds.[24–26] The most popular methods are free-energy perturbation (FEP) and thermodynamic integration (TI).

The free energy difference between two ligands, A and B, that bind to the same pocket can be determined by considering the alchemical transformation A → B int a number of successive steps. This is accomplished by defining a coupling parameter (λ) that controls the smooth change between initial ($\lambda = 0$) and final ($\lambda = 1$) states, so that at each intermediate step one can define the Hamiltonian $H(\lambda)$ as follows:

$$H(\lambda) = \lambda H_B + (1 - \lambda) H_A \qquad (4.1)$$

where H_A denotes the Hamiltonian for the system with ligand A.

In FEP calculations, the free energy change for the transformation of ligand A into B, either free in solution or in the protein–ligand complex, can be determined by the addition of the free energy changes for each window leading from initial to final states:

$$\Delta G = -\sum_{\lambda} RT \ln \langle e^{-\Delta H_\lambda / RT} \rangle_\lambda \qquad (4.2)$$

where $\Delta H_\lambda = H_{\lambda + \Delta \lambda} - H_\lambda$.

TI provides an alternative solution, where the change in free energy can be determined as indicated in eqn (4.3), where one has to evaluate the ensemble average of the derivative of the Hamiltonian with respect to the coupling parameter λ.

$$\Delta G = \int_0^1 \langle \frac{\partial H}{\partial \lambda} \rangle_\lambda \, d\lambda \qquad (4.3)$$

The prediction of the relative binding free energies between two ligands can be related to specific chemical modifications introduced in the ligand. As noted in Figure 4.4, this can be determined by alchemical mutations that convert the two ligands in the unbound and bound states, allowing for extensive sampling of the protein–ligand complex and of the free ligand (in solution) in realistic environments [eqn (4.4)].

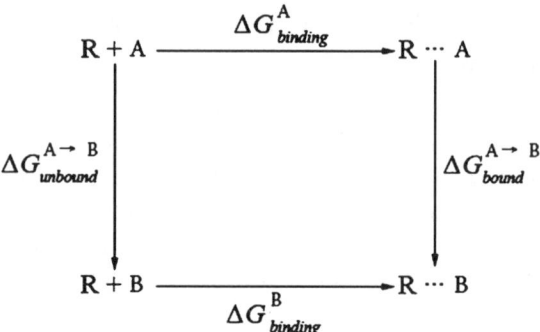

Figure 4.4   Schematic representation of the thermodynamic cycle used for the calculation of relative binding affinities between ligands A and B binding to a common target.

$$\Delta\Delta G = \Delta G_{\text{binding}}^{\text{B}} - \Delta G_{\text{binding}}^{\text{A}} = \Delta G_{\text{bound}}^{\text{A}\rightarrow\text{B}} - \Delta G_{\text{unbound}}^{\text{A}\rightarrow\text{B}} \qquad (4.4)$$

4.2.2   Exploring Pathways and Mechanism of Drug Binding

MD simulations also provide an efficient way to generate unbiased structures to be used in docking or virtual screening analyses.[27,28] Thus, MD simulations can be used not only for the refinement of docked complexes, but also during the preparation of the protein receptor model before docking, in order to optimize the structure and account for protein flexibility (see, for instance, refs 29–31). This is of particular relevance for studies dealing with membrane-bound proteins due to technical complexities associated with crystallization for X-ray analysis and solubilization for NMR analysis.[32,33]

The suitability of atomistic MD techniques to characterize slower motions such as bending of domains through hinge regions or allosterism is more questionable, as the time scales of those structural changes generally demand extensive conformational sampling, which in turn requires a huge amount of computational effort *via* conventional MD techniques. Nevertheless, recent studies have succeeded in gaining insights into the mechanistic details of ligand binding using unbiased MD simulations.

Dror *et al.* examined the process by which drug molecules spontaneously associate with G protein-coupled receptors to achieve poses that match the arrangements found experimentally for the ligand–target protein.[34] By running 82 simulations lasting from 1 to 19 μs, they examined the binding of four distinct ligands – three antagonists (propranolol, alprenolol and dihydroalprenolol) and one agonist (isoproterenol) – to the β_2-adrenergic receptor, and also the binding of dihydroalprenolol to the β_1-adrenergic receptor, starting from systems where ligands were positioned at least 30 Å from the binding pocket and 12 Å from the receptor surface. Interestingly, the analysis of the simulations shed light on a dominant binding pathway comprising two major steps: first, association with the extracellular vestibule, which involves a substantial dehydration and second displacement from the extracellular vestibule into the binding pocket through a narrow passage and adoption of transient poses before relaxing into the crystallographic one.

An independent study by Buch *et al.* examined the formation of an enzyme–inhibitor complex by MD simulations.[35] In this work, the authors reconstructed the complete binding process of the benzamidine–trypsin complex by performing 495 simulations (100 ns each) of free ligand binding. They found that 187 simulations led to binding events with an RMSD (root-mean square deviation) of less than 2 Å compared with the crystal structure. Moreover, the analysis was useful for characterizing the binding pathway, which involves five different metastable states. The estimated binding affinity (-5.2 ± 0.4 kcal mol^{-1}) was in good agreement with the experimentally measured value of -6.2 kcal mol^{-1}.

Such an unbiased 'brute-force' approach is limited by the time scale of the event under study. In the case of NR, studying the binding of ligands in the internal cavity would be impractical, because it involves a slow diffusion process through the protein matrix. However, it is applicable to the study of molecules binding on superficial cavities, such as AF-2 or other allosteric pockets [*e.g.*, the BF-3 described for AR (see, for instance, ref. 36)].

4.2.3 Enhanced Sampling Techniques

The preceding examples suffice to demonstrate the predictive power of MD simulations. Nevertheless, the routine application of the preceding studies is severely handicapped by the large computational cost required to obtain a proper sampling of the system. Therefore, one has to resort to enhanced sampling techniques in order to force the correct sampling of the pathways and mechanisms involved in protein–ligand interactions.

The use of simplified (coarse-grained) representations of the molecular system can be a valuable strategy for enhancing sampling of 'slow' motions by modeling the system as a collection of beads that integrate 'fast' degrees of freedom.[37] As the number of beads decreases, the less expensive is the simulation and the larger the system that can be simulated, thus facilitating the sampling of large structural changes. However, the development of an accurate force field capable of describing the general dynamics of systems becomes more challenging as the graining is 'coarser'. On the other hand, while these techniques can be informative about the general trends of protein dynamics, the absence of atomistic details might limit the identification of structure–function relationships.

Other techniques have been conceived for enhancing conformational sampling through modification of the conventional MD sampling, so that the system is forced to explore the conformational space by facilitating the escape from local energy minimum wells by using non-Boltzmann sampling. In the context of binding/unbinding processes, they often require *a priori* knowledge about the nature of the bound and unbound states, and also of the structural parameters that mediate the transition between them (*i.e.*, the reaction coordinate). As an exhaustive description of these techniques is beyond the scope of this work, we limit ourselves to describing the essential details of a selected set of computational strategies, whose range of applications will be illustrated below by a few representative studies.

4.2.3.1 Replica Exchange Methods

This technique was developed with the aim of enhancing conformational transitions by crossing energy barriers in the rugged energy landscape of biomolecules.[38,39] Replica exchange (RE) involves running a number of simulations at different temperatures and exchanging the temperature and coordinates every fixed number of steps based on Metropolis criteria. Therefore, one can obtain an improved sampling by exchanging complete configurations and allowing low-temperature systems to access a representative set of low-temperature regions of phase space. Due to the simulation of multiple replicas, RE requires a proportionally higher computational effort. Moreover, the set of temperatures should be chosen to ensure that no replica becomes trapped in local minima, and also to achieve proper swapping of adjacent replicas.

Different RE variants have been reported in the literature. For instance, Berne and co-workers developed an RE method with solute tempering variant able to include explicitly the solvent contribution, but also to reduce the number of replicas by avoiding the evaluation of solvent–solvent interactions.[40] Another variant consists of investigating a set of different modified potentials, as implemented in Hamiltonian REMD (replica exchange molecular dynamics).[41,42] Finally, RE techniques can been combined with FEP and TI calculations in order to estimate relative free energies of systems involving large conformational reorganizations.[43]

4.2.3.2 Targeted Molecular Dynamics

Targeted molecular dynamics (TMD) is used to drive the system from an initial state to a final 'target' state by using time-dependent holonomic constraints.[44] To this end, a subset of atoms in the system is guided towards a final 'target' structure by means of steering forces. At each time step, the simulated system is first aligned to the target state and then the RMSD between the current coordinates of the simulated system and the corresponding atoms in the target structure is computed. The potential energy is then modified by the addition of a correction term that follows the expression

$$\Delta V = \frac{1}{2}\frac{k}{N}[\text{RMSD}(t) - \text{RMSD}*(t)]^2 \qquad (4.5)$$

where $\text{RMSD}(t)$ denotes the instantaneous RMSD from the current structure to the target structure, $\text{RMSD}*(t)$ varies linearly from the initial RMSD at the first step to the final value at the last step of the simulation, k is the force constant and N denotes the number of targeted atoms.

4.2.3.3 Random Acceleration Molecular Dynamics

A method proposed to investigate plausible pathways for egression of ligands is random acceleration molecular dynamics (RAMD). In this technique, an artificial force with random direction is applied to the centre of mass of one molecule in the simulated system (*i.e.*, the ligand) in addition to the standard classical force field, as was originally designed to enhance the probability of substrates to exit from the buried active site of proteins.[45] The direction of the additional force is chosen randomly and is kept for a chosen number of time steps (N). During this time period, a specified distance should be covered by the substrate, which has to travel at an average velocity during the time period $N \times \Delta t$, where Δt is the time step of the MD simulation. If the substrate encounters rigid parts of the cavity, its average velocity will fall below the pre-set threshold. In this case, a new direction is chosen randomly and maintained as long as the substrate moves in the new direction with an average velocity greater than the pre-set value. This scheme permits the substrate to explore different regions of the binding pocket until it finds an exit pathway.

4.2.3.4 Steered Molecular Dynamics

Steered molecular dynamics (SMD) is a non-equilibrium sampling technique. In SMD calculations, an external force is applied to the system in order to promote the variation of a suitable variable, such as the distance of the ligand from a reference point, in a relatively fast manner. The external force $F(t)$ is often modeled as a spring connected to a dummy atom, which drives the system to the desired state at constant velocity [eqn (4.6)]

$$F(t) = k(x_0 + vt - x) \tag{4.6}$$

where x denotes a given geometric parameter (*i.e.*, a distance, angle or torsion), x_0 is the initial position of the reference atom and v is the constant velocity that drives the time evolution of the atom.

The free energy differences can then be determined by using Jarzynski's equality [eqn (4.7)],[46] which relates equilibrium free energy values to the irreversible work (W) performed over the system, proceeding along a reaction coordinate that connects initial and final states (*i.e.*, bound and unbound ligand).

$$\exp(-\Delta F / k_B T) = \langle \exp(-W / k_B T) \rangle \tag{4.7}$$

4.2.3.5   Metadynamics

An alternative strategy is metadynamics, where the evolution of the system is biased by adding a history-dependent potential energy function of a determined set of collective variables to the Hamiltonian of the system.[47] The potential is constructed through successive addition of Gaussian functions that act as repulsive potentials to prevent the system from revisiting points of the free energy surface [eqn (4.8)]. Consequently, the system can escape the wells in the rugged landscape and efficiently explore it. The sum of Gaussian functions deposed along the trajectory of the system is then used to reconstruct the free energy

$$\Delta V(\xi,t) = \int_0^t \frac{\omega}{\tau_G} \exp\left(- \frac{\{\xi[r(t)] - \xi[r(\tau)]\}^2}{2\sigma} \right) d\tau \qquad (4.8)$$

where ξ represents a reaction coordinate, ω denotes the height of the Gaussian and τ_G is the deposition stride.

The reaction coordinate is composed of variables that are functions of the coordinates of the system well suited for describing its motion and dynamics. Ideally, the collective variables should distinguish initial, final and intermediate states and describe all the slow events that are relevant to the process of interest. Unfortunately, there is no *a priori* recipe for finding the correct set of collective variables and often a trial-and-error process is required. The simplest definition of these variables involves geometry-related variables, but more elaborate variables, such as normal modes and essential coordinates derived from essential dynamics, are also valid.[48–50] An interesting recent development of this technique is the use of self-learning algorithms able to define the reaction coordinate automatically.[51,52]

4.3   Ligand Binding to Nuclear Receptors

In recent years, a significant number of studies have shown the relevance of computational methods for understanding the mechanisms of ligand binding/unbinding and the dynamics of NRs. The increasing number of studies dealing with computational analysis of these molecular mechanisms is clearly beyond the scope of this chapter. Accordingly, just a representative selection of the most recent studies are discussed to illustrate the potential impact of computational tools.

4.3.1   Ligand Recognition and Specificity

The information gained from computational studies on the features that mediate the specific recognition of ligands is illustrated by recent work reported by Alvarez *et al.* on the binding mode of dafachronic acid (DA) and structurally related compounds to the LBD of the DAF-12 receptor.[53] The dynamic behavior of the DAF-12 complex with $(25S)$-Δ^7-DA revealed two distinct binding modes, one of them being energetically more favorable compared with the 25R isomer. TI was then used to determine the relative free energies of $(25R)$- and $(25S)$-Δ^7-DA. The results indicated that the introduction of a methyl group in both C-25 configurations increases the ligand affinity. However, whereas for the R configuration this gain is only 0.9 kcal mol^{-1}, for the S configuration it is almost twice that value (1.7 kcal mol^{-1}). Therefore, the binding energy of $(25S)$-Δ^7-DA is 0.8 kcal mol^{-1} stronger than the affinity of $(25R)$-Δ^7-DA, which is consistent with the higher activity reported for the 25S stereoisomer. The analysis of the interactions suggested that the higher activity of $(25S)$-Δ^7-DA could be related to the strong electrostatic interaction produced in the fully extended conformation of the $(25S)$ isomer side chain. Moreover, since this side chain is more flexible within the ligand-binding pocket (LBP), the entropic penalty due to binding should be lower than in the more rigid 25R isomer side chain, increasing its affinity.

Selectivity of ligand binding has been examined in recent studies for thyroid hormone (TH) receptor (TR) and the constitutive androstane receptor (CAR). TH receptors influence important physiological processes related to alert state, such as heart frequency and metabolic rates. Whereas TRβ has beneficial effects by lowering cholesterol levels, TRα mediates harmful effects on the heart.[54,55] On the other hand, in addition to being involved in the metabolic conversion of heme, bilirubin, bile acids and TH, CAR might be a metabolic sensor, as it can recognize toxic endogenous metabolic by-products and also exogenous chemicals and the interaction with ligands induces the expression of metabolizing enzymes and transporter proteins involved in the elimination of potentially toxic chemicals.[56]

Martinez *et al.* investigated the behavior of Triac (3,5,3′-triiodothyroacetic acid), a TR agonist binding 2–3-fold selectively to TRβ with respect to TRα.[57] This preference is not supported by inspection of X-ray structures, which suggests that Triac better fits the TRα LBD. The ligand carboxylic moiety interacts with the backbone of Ser277α and the side chain of Arg266α and the 4′-hydroxyl group contacts His381α. Further stabilization is afforded by two water-mediated contacts with Ser277α and Arg228α. In contrast, the complex with TRβ only involves hydrogen bonds between the ligand hydroxyl moiety and the side chain of His435 and between the carboxylic group and the backbone of Asn331.

MD simulations were performed to investigate the origin of ligand selectivity. Ligand–receptor interaction energies confirmed the impression that TRα makes stronger contacts with the ligand, but similar values were obtained for TRα and TRβ when the whole system (*i.e.*, protein, ligand and water) was considered, indicating that favorable interactions with water molecules in TRβ compensate for weaker ligand–protein contacts. On the other hand, the presence of several waters at the complex interface increases the ligand mobility and thus the ligand entropy. In fact, the conformational entropy gain of Triac in TRβ *versus* TRα was estimated to be 1.5 cal K^{-1}

mol^{-1}, showing that entropy plays a fundamental role in protein–ligand recognition.[58–60] Analogously, the improved affinity of some vitamin D receptor ligands has been associated with multiple ligand conformations, increased LBD volume and waters in the LBD binding pocket,[61,62] and the estrogen receptor β (ERβ) LBD was expanded when complexed with THC, a selective agonist binding to ERβ with four-fold higher affinity with respect to ERα which is, instead, antagonized.[63]

These findings suggest that it would be interesting to evaluate other NR complexes by exploiting an enthalpy/entropy compensation strategy rather than only increasing the complementarity of ligands towards the binding site. In particular, the reduction of the size of ligands occupying flexible regions of the LBD could represent an innovative and promising strategy for increasing entropy and, thus, for ameliorating the binding free energy.

Windshügel and Poso carried out MD simulations to investigate the molecular basis for constitutive activity and ligand-dependent receptor activation of CAR.[64] It is widely believed that the receptor assumes an active conformation even in the absence of the natural substrate, but unequivocal structural evidence is still lacking.[65–67] To unravel the molecular basis of the constitutive activity of CAR and to investigate the effect of agonist binding, the authors examined two structures in the presence and absence of two co-crystallized ligands, pregnanedione and CITCO {6-(4-chlorophenyl)imi-dazo[2,1-*b*][1,3]thiazole-5-carbaldehyde *O*-(3,4-dichlorobenzyl)oxime}, by means of MD simulations. In contrast to other NRs, CAR possesses an additional helix, known as HX, which seems to be involved in the receptor constitutive activity, as it maintains H12 in the active conformation even in the absence of ligand (Figure 4.5). The calculations helped to model the position of HX in the native protein, since no apo crystallographic structure is currently available. Moreover, they underlined how the ligand binding restricts the conformational flexibility of LBD and in particular of the H1–H3 loop, the H6/H7 region and the loop connecting H9 and H10/H11.

The basal activity of the receptor was supported by the formation and maintenance during the trajectories of the uncomplexed forms of several hydrogen bonds. Tyr326 contacts Cys347 on H12, thus resembling the agonist-H12 contact experienced by other NR/agonist complexes. H12 is also contacted by Gln331 and an additional backbone–backbone interaction between Gly196 and Ile346 (not present in the crystallographic structures) was observed. Van der Waals contacts were, instead, less conserved in ligand-free simulations, although their importance was confirmed for ligand-bound simulations, which showed the interaction of Phe161 with Leu336 located on HX and with Met340 placed between HX and H12. The van der Waals interactions are increased upon binding of the two agonists to the LBD pocket, limiting the receptor flexibility. Thus, the presence of an agonist preserves the helical shape and stabilizes the conformation of H12, which only experiences partial unwinding in ligand-free simulations.

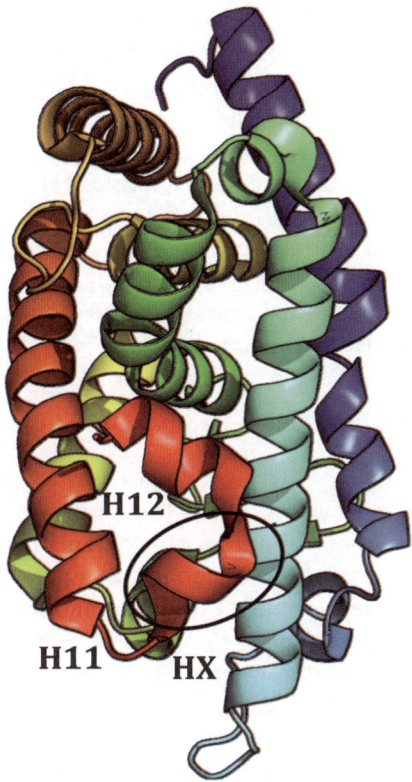

Figure 4.5   Representation of helix HX in the structure of the constitutive androstane receptor (PDB code 1XNX).

On the basis of these observations, the authors proposed the standard model of ligand-mediated NR activation, where ligands modulate the H12 stabilization. The model considers that the binding of an agonist shifts Phe161 towards the LBD/H12 interface, leads to new hydrogen bonds and restricts the conformational flexibility of helices H11 and HX. At the same time, the interactions of the ligand with Tyr326 reorient the residue side chain, which contacts H12 preserving the helix folding and location. Overall, the simulations provided clues for a better understanding of ligand-dependent activation of xenosensors.

Finally, in a recent study, Jyrkkärinne *et al.* investigated the movements of CAR H12 in the presence of both activating and repressing ligands and of a co-repressor peptide.[68] Particular attention was paid to understanding inverse agonism (since CAR H12 maintains the same closed active conformation in the presence or absence of any molecule, ligand binding may, in fact, result in agonism or inverse agonism). To this end, the authors docked seven agonists and eight inverse agonists into CAR LBD and the complexes were simulated in the presence of a co-repressor peptide. The results indicated that the presence of an inverse agonist destabilized HX with respect to apo or to agonist-bound structures. In contrast to other NRs, CAR H12 did not experience a large translocation upon binding of co-repressors or inverse agonists. Then, any disruption of HX structure might lead to a similar destabilization of H12 by displacing it from the active position. This adjustment is induced by the co-repressor peptide, whose accommodation relocates H12. The subtle movement of H12 towards H10 produced by inverse agonists and testified by changes in the van der Waals interactions between the two helices might, thus, be sufficient for co-repressor binding, as already suggested by Küblbeck *et al.*[69] Overall, these results facilitated the identification of major structural differences between agonist- and inverse agonist-bound complexes.

4.3.2 Ligand Unbinding

The identification of ligand entry/exit routes is fundamental for a proper understanding of protein function, and also to provide guidelines in rational drug design. This information cannot be gained in a straightforward way from experimental studies, but exploration of ligand migration pathways is well suited to MD-based simulation tools. Ligand unbinding in NRs is, however, a controversial yet unsolved challenge. For instance, while it was suggested that detachment of H12 is necessary for ligand unbinding from the LBD pocket in the retinoic acid receptor (RAR),[70] SMD simulations identified two distinct pathways, one for ligand binding and a 'back door' for ligand unbinding,[71] and random expulsion MD suggested that the retinoic acid may exit the binding site through flexible regions close to the H1–H3 loop and β-sheets, without displacing H12 from its agonist position.[72] The studies performed by Polikarpov and co-workers on TR also indicated that ligand unbinding does not necessarily imply the displacement of H12.[73–75] Up to four different routes were found for the ligand exiting from the ER LBD, two based on the H12 dislocation and other two on the separation of H8 and H11.[76]

Shen *et al.* explored the unbinding pathways followed by the two selective ERβ ligands WAY-244 and genistein in both ERα and ERβ and related them to ligand selectivity.[77] Four systems, *i.e.*, the ERα and ERβ isoforms complexed with WAY-244 and genistein, were first equilibrated through short MD simulations and then subjected to SMD simulations. The four pathways analyzed were (Figure 4.6) (i) the area between the H11–H12 loop and the terminal part of H3 (*i.e.*, the mousetrap mechanism), (ii) the displacement of H12, (iii) the region between the H7–H8 loop and H11 and (iv) the region between H7 and the H2–H3 loop involving the H7 unfolding.

The results revealed that the third route, passing among the H7–H8 loop and H11, is the most favorable unbinding pathway, since the work required for pulling out the ligand was definitely smaller than that for the other cases. The analyses also confirmed the fundamental role played by His524 in ERα and His475 in ERβ, both acting as gatekeepers. The bottleneck for exiting processes was, in fact, represented by the clashes between the ligands and the

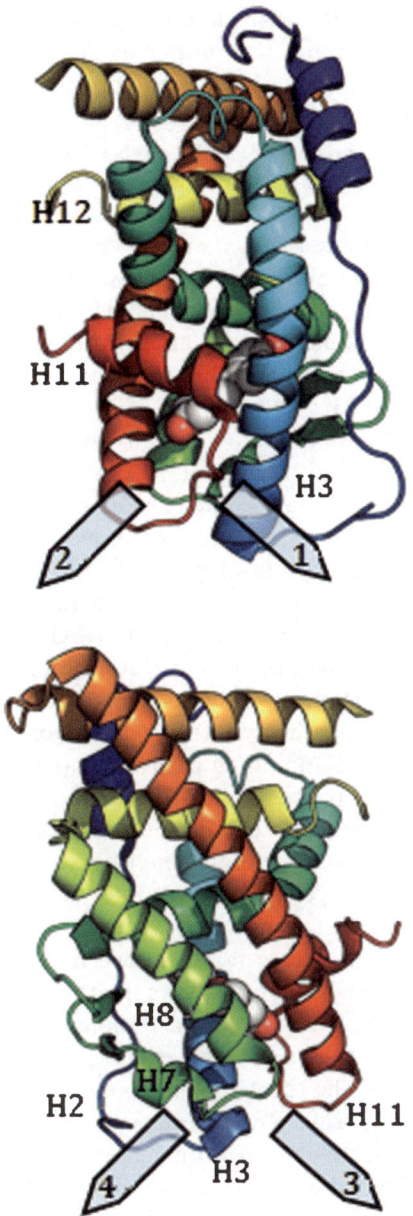

Figure 4.6 Representation of the four main pathways of ligand unbinding found for estrogen receptor. The structure of ERα complexed with the ligand WAY-244 is reported as reference structure (PDB code 1X7E).

side chains of residues, in agreement with previous studies.[78,79] Finally, the authors observed how the different behavior of ligands in unbinding from ERα and ERβ could be exploited for understanding and increasing ligand selectivity. In particular, in the ERβ case, the polar region at the H7–H8 loop C-terminal part acted as a polar transmitter in the genistein unbinding, since the polar groups of the ligand form a number of hydrogen bonds that counterbalance the steric clashes. Accordingly, enhancing selectivity towards ERβ might be modulated by increasing the binding affinity to ERβ and decreasing the potential of mean force (PMF) for ligand unbinding from ERα or by reducing the binding affinity to ERα and increasing the PMF for ligand unbinding from ERβ. Since routes exiting ER receptors are essentially hydrophobic and ERβ unbinding was demonstrated to be regulated by the H7–H8 loop transmitter, the authors concluded that a successful strategy could be represented by reducing the polar substituents on the dual-ring part or by adding polar functions on the phenolic moiety of the ligand.

Ligand unbinding from ERα and ERβ was also investigated by Burendahl *et al.*[80] It is well known that NR activation involves the interaction with co-activators at the AF-2 site and that ligand release is somehow followed by co-activator dissociation.[81] For the specific case of ERα, an additional cofactor peptide site was also identified in the region above the β-sheet, known as the αII site.[82,83] Thus, the authors considered both ERα and ERβ isoforms, the natural ligand 17β-estradiol, the agonist genistein, the antagonist 4-hydroxytamoxifen and the cofactor peptides binding to AF-2 and αII sites. RAMD and SMD techniques were then used to explore the unbinding pathways.

The search for egression pathways was performed on six different LBD models, but of the total pathways only some were identified in all the simulation. The main ligand unbinding pathway was located between the H1–H3 loop, the H11–H12 loop and the cleft between H7 and H11. The second was found close to the β-sheet and the H2 loop and the third between H11, the H11–H12 loop and H12. The fourth and the fifth were found exiting between H6–H7 and H2–H3, respectively, the sixth under H4 and, finally, the seventh passed in the narrow area between H3, H4 and the AF-2 cofactor.

Although some differences were found between the results obtained from RAMD and SMD methods, the analysis suggests that ligand unbinding can occur without inducing significant conformational changes in the receptor. The ligand can escape the LBD by exploiting subtle local side-chain rearrangements, whereas the protein backbone only experiences thermal fluctuations. Nevertheless, since the exiting is a dynamic process associated with conformational adjustments in both ligand and receptor, the chemical features of the ligand likely dictate the choice of the unbinding pathway. Thus, RAMD simulations performed with the bulkier antagonist 4-hydroxytamoxifen did not lead to any ligand exiting, suggesting that antagonist unbinding requires larger conformational changes. Although these findings might agree with previous suggestions about the involvement of distinct mechanisms for unbinding of agonists and antagonists,[84] a larger conformational sampling seems necessary as the unbinding of the antagonist.

A last example comes from a study by Peräkylä,[85] who used a variety of computational techniques to investigate the unbinding of 1α,25-dihydroxyvitamin D_3 (VD3) from the ligand-binding pocket of the vitamin D receptor (VDR). Peräkylä combined a total of 200 RAMD trajectories for the unbinding of VD3 from the VDR LBP with a subsequent 40 reference TMD calculations, which yielded ligand unbinding trajectories varying from 0.5 to 2.5 ns, and finally characterization of energy profiles by SMD simulations. The preceding strategy led to the identification of up to five main exit pathways, with two of them (pathways A and B) exhibiting the highest frequencies for ligand unbinding. In pathway A, the ligand exit occurs through a channel created by rearrangement of H12. In contrast, pathway B is located in the opposite direction of the LBD compared with pathway A, involving an opening at the protein surface between the H1–H2 loop and the short β-sheet between H5 and H6. Compared with other studies that identified another pathway involving H3 adjustment in TRβ,[73] a pathway located between the H1–H2 loop and H3 was found to have a minor contribution in RAMD simulations, but a higher weight in TMD simulations. Finally, on the basis of SMD simulations, pathway B seemed to be the most favourable for VD3 unbinding from the VDR LBP.

Summing up, the preceding discussion confirms that ligand unbinding is a dynamic process that involves distinct pathways depending on the chemical nature of the ligand and the specific structural features of the receptor. Moreover, it is worth noting that the results should be considered in the framework of a multiprotein complex, which possibly occludes some of the identified routes. At this point, recently published structures of NR PPAR-γ complexed with RXR-α and DNA,[86] and of LRH-1 with the cofactor DAX1,[87] showed how the interaction regions overlie the exit of a number of the identified routes. It follows that unbinding processes could be strongly affected by the interaction with DNA and cofactors and also by the tissue-dependent nature of the different cofactors, thus stressing again the necessity for performing new investigations including all the different players and full-length nuclear receptors.

4.3.3 Allosterism

Allosterism is a fundamental property for modulating the biological function of proteins. In the case of NRs, the formation of the receptor–ligand complex directs the transcription of genes in a process that is assisted by recruitment of co-regulators and abnormalities of co-activator binding and related gene expression have been related to several disorders.[88] Accordingly, there is great interest in deciphering the molecular basis of co-regulator binding and its influence on gene transcription.

In the case of the androgen receptor (AR), the co-activator proteins have been classified broadly into those containing FxxLF and those containing LxxLL motifs.[89,90] The interaction of co-activator with AR has recently been shown to be a biphasic process involving an initial association followed by conformational rearrangements.[91] Moreover, these studies indicate that each co-activator induces a distinct conformational state in the dihydrotestosterone:AR-LBD:co-activator complex, which involves different intramolecular rearrangements in the AR-LBD backbone. Thus, even in the presence of the same ligand, AR-LBD can adopt distinct conformational states depending on its interactions with specific co-activators.

Molecular simulations provide a promising strategy to gain insight into the molecular mechanisms that mediate the action of co-regulators on the activity of nuclear receptors. This is illustrated by a recent work by Cunningham on liver X receptor (LXR).[92] LXR forms heterodimers with 9-*cis*-retinoic acid receptors (RXR).[93] The LXR monomer forms a bundle of 10 α-helices, whereas the RXR monomer forms a bundle of 11 α-helices. The LXR–RXR dimer can be activated exclusively by an RXR agonist, by an LXR agonist or by both. The permissive action of the LXRs suggests communication pathways in the heterodimer complex often described as allosteric effects. The allosteric network connects four structurally distinct but functionally coupled regions of the RXR heterodimer complex: the ligand binding site, the cofactor binding surface, the AF-2 site and the dimerization surface between the receptors.

The effect of ligand binding in the heterodimer and the transfer of signal transduction through the dimer interface in this system was studied by performing extended MD simulations on the heterodimer with a ligand bound to each monomer, on the dimer with a ligand bound to a single monomer in the heterodimer, and finally without any bound ligand. The change in the centre-of-mass position for each residue was then analyzed by means of cross-correlation techniques, showing that most of the interactions between secondary structures of each of the LXR and RXR subunits are strongly persistent. In contrast, correlations between the two helices that define the dimer interface exhibit significant alterations related to the presence or absence of ligands. As an example, the interaction formed between Arg401 of LXR and Glu465 of RXR when the two ligands are bound is replaced by the interaction of Arg401 with Glu472 of RXR when the LXR ligand is absent. Overall, binding of ligands changes the network of interactions between the helices, in agreement with previous studies.[94] Moreover, these results suggest that the structural changes observed in MD simulations and the corresponding changes in the interaction patterns, can provide a hint that, at much longer time scales, a conformational shift, mediating the allosteric control, may arise and promote the action of NRs.

In addition to the effect of ligands on the signaling between monomers in the heterodimer, another aspect of nuclear receptor activation concerns the interaction with cofactors. This issue was addressed by Burendahl and Nilsson,[95] who performed MD simulations of the LXR–RXR heterodimer with nine different cofactor peptides, including both co-activator and co-repressor peptides, with the aim of identifying specific interactions relevant for the binding affinity and for signal transduction. The results indicated that the cofactor peptide–LXR recognition generally involves a strong clamp interaction, often supplemented by specific secondary contacts. In other instances, there is a weak charge clamp interaction compensated by other strong interactions. More interestingly, the interaction patters indicate that the co-activator and co-repressor peptides interact with the receptor in similar ways, although a larger fluctuation was found for residues in the N-terminal region of the peptide in the case of co-activators. Finally, the signal transduction from the peptide cofactor to the ligand in the LXR LBP is mediated by a leucine located at the N-terminal end of the LxxLL motif, which appears to be a key signaling residue and signal transmission to LXR is mediated by Leu290 and Lys291 on H4. After Lys291, the signal can take two different routes, either passing over to H12 Ile442 and Trp443 or continuing to H5 through the backbone. Along H5, several interaction points with the ligand can be identified.

Overall, the above study suggests that signaling pathways involve transmission both through the backbone and through van der Waals contacts between side chains. Whereas backbone signaling can be obtained in a sequence-independent fashion, side-chain signaling is sequence dependent. Both the extent of the energetic couplings and its transmission pathway can then be used to tune the receptor response to the unique molecular interactions of each complex.

4.4 Final Remarks

A survey of recent research accounts clearly indicates that the application of MD techniques offers excellent prospects to deepen our knowledge of the biological role of NRs. On the one hand, general methodological and computational developments have taken the field of MD to new heights: refined force fields produce conformational ensembles of equivalent quality to experimental ones,[13] adaptation of MD programs to new hardware permits simulations for unprecedented lengths of time, reaching the millisecond scale,[96,97] and enhanced sampling techniques and MD-coupled free-energy methods have demonstrated their usefulness and keep on improving. On the other hand, the structural biology of NRs – underpinned by a wealth of X-ray crystallographic data – is moving towards a dynamic and more complete view of the systems that aims to understand the mechanisms of ligand binding, functional regulation or interplay with other biological components. Thanks to the uniquely detailed description of the molecular phenomena, MD simulations are essential for obtaining a detailed understanding of the biology of NRs.

Acknowledgements

This work was supported by the Spanish Ministerio de Innovación y Ciencia (SAF2011-27642, SAF2009-08811) and the Generalitat de Catalunya (2009-SGR00298).

References

1. Y. Brélivet, N. Rochel and D. Moras, *Mol. Cell. Endocrinol.*, 2012, **30**, 466.
2. E. Hur, S. J. Pfaff, E. Sturgis Payne, H. Grøn, B. M. Buehrer and R. J. Fletterick, *PLoS Biol.*, 2004, **2**, e274.
3. K. I. Pardee, X. Xu, J. Reinking, A. Schuetz, A. Dong, S. Liu, R. Zhang, J. Tiefenbach, G. Lajoie, A. N. Plotnikov, A. Botchkarev, H. M. Krause and A. Edwards, *PLoS Biol.*, 2009, **7**, e43.
4. E. J. Woo, D. G. Jeong, M. Y. Lim, S. J. Kim, K. J. Kim, S. M. Yoon, B. C. Park and S. E. Ryu, *J. Mol. Biol.*, 2007, **373**, 735.
5. J. Lihua and L. Yong, *Adv. Drug Deliv. Rev.*, 2010, **62**, 1218.
6. M. Nocker and P. Cozzini, *Curr. Top. Med. Chem.*, 2011, **11**, 133.
7. A. Le Maire, C. Teyssuer, C. Erb, M. Grimaldi, S. Alvarez, A. R. de Lera, P. Balaguer, H. Gronemeyer, C. A. Royer, P. Germain and W. Bourguet, *Nat. Struct. Mol. Biol.*, 2010, **17**, 801.
8. W. Bourguet, V. Vivat, J. M. Wurtz, P. Chambon, H. Gronemeyer and D. Moras, *Mol. Cell*, 2000, **5**, 289.
9. S. A. Adcock and J. McCammon, *Chem. Rev.*, 2006, **106**, 1589.
10. M. Karplus and J. Kuriyan, *Proc. Natl. Acad. Sci. U. S. A.*, 2005, **102**, 6679.
11. G. Dodson and C. S. Verma, *Cell Mol. Life Sci.*, 2006, **63**, 207.
12. A. Jain and G. Stock, *J. Chem. Theory Comput.*, 2012, in press (DOI: 10.1021/ct300077q).
13. K. A. Beauchamp, Y.-S. Lin, R. Das and V. S. Pande, *J. Chem. Theory Comput.*, 2012, **8**, 1409.
14. H. Nury, F. Poitevin, C. van Renterghem, J. P. Changeux, P. J. Corringer, M. delarue and M. Baaden, *Proc. Natl. Acad. Sci. U. S. A.*, 2010, **107**, 6275.
15. P. Fredolino, M. Gruebele and K. Schulten, *Biophys. J.*, 2008, **94**, L75.
16. E. A. Cino, J. Wong-ekkabut, M. Karttunen and W.-Y. Choy, *PLoS One*, 2011, **6**, e27371.
17. P. Maragakis, K. Lindorff-Larsen, M. P. Eastwood, R. O. Dror, J. L. Klepeis, I. T. Arkin, M. O. Jensen, H. Xu, N. Trbovic, R. A. Friesner, A. G. Palmer III and D. E. Shaw, *J. Phys. Chem. B*, 2008, **112**, 6155.
18. M. J. Harvey, G. Giupponi and G. de Fabrittis, *J. Chem. Theory Comput.*, 2009, **5**, 1632.
19. A. Tomita, T. Sato, K. Ichiyanagi, S. Nozawa, H. Ichikawa, M. Chollet, F. Kawai, S. Y. Park, T. Tsuduki, T. Yamato, S. Y. Koshihara and S. Adachi, *Proc. Natl. Acad. Sci. U. S. A.*, 2009, **106**, 2612.
20. M. Billeter, G. Wagner and K. Wuthrich, *J. Biomol. NMR*, 2008, **42**, 155.
21. J. R. Allison, P. Varnai, C. M. Dobson and M. Vendruscolo, *J. Am. Chem. Soc.*, 2009, **131**, 18314.
22. M. M. Dedmon, K. Lindorff-Larsen, J. Christodoulou, M. Vendruscolo and C. M. Dobson, *J. Am. Chem. Soc.*, 2005, **127**, 476.
23. C. Camilloni, P. Robustelli, A. de Simone, A. Cavalli and M. Vendruscolo, *J. Am. Chem. Soc.*, 2012, **134**, 3968.

24. C. Chipot and A. Pohorille (eds), *Free Energy Calculations. Theory and Applications in Chemistry and Biology*, Springer Series in Chemical Physics, Vol. 86, Springer, Berlin, 2007.
25. A. de Ruiter and C. Oostenbrink, *Curr. Opin. Chem. Biol.*, 2011, **15**, 547.
26. J. D. Chodera, D. L. Mobley, M. R. Shirts, R. W. Dixon, K. Branson and V. S. Pande, *Curr. Opin. Struct. Biol.*, 2011, **21**, 150.
27. J. Kua, Y. Zhang and J. A. McCammon, *J. Am. Chem. Soc.*,2002, **124**, 8260.
28. J. H. Lin, A. L. Perryman, J. R. Schames and J. A. McCammon, *J. Am. Chem. Soc.*, 2002, **124**, 5632.
29. H. Alonso, A. A. Bliznyuk and J. E. Gready, *Med. Res. Rev.*, 2006, **26**, 531.
30. K. L. Meagher and H. A. Carlson, *J. Am. Chem. Soc.*, 2004, **126**, 13276.
31. E. Novoa, L. Ribas de Pouplana, X. Barril and M. Orozco, *J. Chem. Theory Comput.*, 2010, **8**, 2547.
32. X. Deupi, N. Dolker, M. L. Lopez-Rodriguez, M. Campillo, J. A. Ballesteros and L. Pardo, *Curr. Top. Med. Chem.*, 2007, **7**, 991.
33. F. Fanelli and P. G. de Benedetti, *Chem. Rev.*, 2005, **105**, 3297.
34. R. O. Dror, A. C. Pan, D. H. Arlow, D. W. Borhani, P. Maragakis, Y. Shan, H. Xu and D. E. Shaw, *Proc. Natl. Acad. Sci. U. S. A.*, 2011, **108**, 13118.
35. I. Buch, T. Giorgino and G. de Fabritiis, *Proc. Natl. Acad. Sci. U. S. A.*, 2011, **108**, 10184.
36. E. Estebanez-Perpina, L. A. Arnold, P. Nguyen, E. D. Rodrigues, E. Mar, R. Bateman, P. Pallai, K. M. Shokat, J. D. Baxter, R. K. Guy, P. Webb and R. J. Fletterick, *Proc. Natl. Acad. Sci. U. S. A.*, 2007, **104**, 16074.
37. V. Tozzini, *Curr. Opin. Struct. Biol.*, 2005, **15**, 144.
38. D. J. Earl and M. W. Deem, *Phys. Chem. Chem. Phys.*, 2005, **7**, 3910.
39. A. Liwo, C. Czaplewski, S. Oldziej and H. A. Scheraga, *Curr. Opin. Struct. Biol.*, 2008, **18**, 134.
40. P. Liu, B. Kim, R. A. Friesner and B. J. Berne, *Proc. Natl. Acad. Sci. U. S. A.*, 2005, **102**, 13749.
41. W. Kwak and U. H. Hansmann, *Phys. Rev. Lett.*, 2005, **95**, 138102.
42. J. Hritz and C. Oostenbrink, *J. Chem. Phys.*, 2008, **128**, 144121.
43. C. Woods, J. Essex and M. King, *J. Phys. Chem. B*, 2003, **107**, 13703.
44. J. Schlitter, M. Engels, P. Krüger, E. Jacopby and A. Wollmer, *Mol. Simul.*, 1993, **10**, 291.
45. S. K. Lüdemann, V. Lounnas and R. C. Wade, *J. Mol. Biol.*, 2000, **303**, 797.
46. C. Jarzynski, *Phys. Rev. Lett.*, 1997, **78**, 2690.
47. A. Laio and M. Parrinello, *Proc. Natl. Acad. Sci. U. S. A.*, 2002, **99**, 12562.
48. F. Trudu, D. Donadio and M. Parrinello, M. *Phys. Rev. Lett.*, 2006, **97**, 105701.
49. V. Spiwok, P. Lipovova and B. Kralova, *J. Phys. Chem. B*, 2007, **111**, 3073.
50. S. Piana and A. Laio, *J. Phys. Chem. B*, 2007, **111**, 4553.
51. G. A. Tribello, M. Ceriotti and M. Parrinello, *Proc. Natl. Acad. Sci. U. S. A.*, 2010, **107**, 17509.
52. G. A. Tribello, M. Ceriotti and M. Parrinello, *Proc. Natl. Acad. Sci. U. S. A.*, 2012, **109**, 5196.

53. L. D. Alvarez, P. Arroyo Mañez, D. A. Estrín and G. Burton, *Proteins*, 2012, **80**, 1798.
54. J. D. Baxter and P. Webb, *Nat. Rev. Drug Discov.*, 2009, **8**, 308.
55. P. M. Yen, *Physiol. Rev.*, 2001, **81**, 1097.
56. A. di Masi, E. de Marinis, P. Ascenzi and M. Marino, *Mol. Aspects Med.*, 2009, **30**, 297.
57. L. Martínez, A. S. Nascimento, F. M. Nunes, K. Phillips, R. Aparicio, S. M. Dias, A. C. Figueira, J. H. Lin, P. Nguyen, J. W. Apriletti, F. A. Neves, J. D. Baxter, P. Webb, M. S. Skaf and I. Polikarpov, *Proc. Natl. Acad. Sci. U. S. A.*, 2009, **106**, 20717.
58. P. Cozzini, M. Fornabaio, A. Marabotti, D. J. Abraham, G. E. Kellogg and A. Mozzarelli, *Curr. Med. Chem.*, 2004, **11**, 3093.
59. G. A. Holdgate, A. Tunnicliffe, W. H. Ward, S. A. Weston, G. Rosenbrock, P. T. Barth, I. W. Taylor, R. A. Pauptit and D. Timms, *Biochemistry*, 1997, **36**, 9663.
60. J. E. Ladbury, *Chem. Biol.*, 1996, **3**, 973.
61. C. Carlberg and F. Molnar, *Curr. Top. Med. Chem.*, 2006, **6**, 1243.
62. F. Molnar, M. Perakyla and C. Carlberg, *J. Biol. Chem.*, 2006, **281**, 10516.
63. A. K. Shiau, D. Barstad, J. T. Radek, M. J. Meyers, K. W. Nettles, B. S. Katzenellenbogen, J. A. Katzenellenbogen, D. A. Agard and G. L. Greene, *Nat. Struct. Biol.*, 2002, **9**, 359.
64. B. Windshügel and A. Poso, *J. Mol. Recognit.*, 2011, **24**, 875.
65. I. Dussault, M. Lin, K. Hollister, M. Fan, J. Termini, M. A. Sherman and B. M. Forman, *Mol. Cell. Biol.*, 2002, **22**, 5270.
66. B. Windshügel, J. Jyrkkärinne, A. Poso, P. Honkakoski and W. Sippl, *J. Mol. Model.*, 2005, **11**, 69.
67. R. X. Xu, M. H. Lambert, B. B. Wisely, E. N. Warren, E. E. Weinert, G. M. Waitt, J. D. Williams, J. L. Collins, L. B. Moore, T. M. Willson and J. T. Moore, *Mol. Cell*, 2004, **16**, 919.
68. J. Jyrkkärinne, J. Küblbeck, J. Pulkkinen, P. Honkakoski, R. Laatikainen, A. Poso and T. Laitinen, *J. Chem. Inf. Model.*, 2012, **52**, 457.
69. J. Küblbeck, J. Jyrkkärinne, F. Molnar, T. Kuningas, J. Patel, B. Windshügel, T. Nevalainen, T. Laitinen, W. Sippl, A. Poso and P. Honkakoski, *Mol. Pharmacol.*, 2011, **8**, 2424.
70. A. Blondel, J. P. Renaud, S. Fischer, D. Moras and M. Karplus, *J. Mol. Biol.*, 1999, **291**, 101.
71. D. Kosztin, S. Izrailev and K. Schulten, *Biophys. J.*, 1999, **76**, 188.
72. P. Carlsson, S. Burendahl and L. Nilsson, *Biophys. J.*, 2006, **91**, 3151.
73. L. Martinez, M. T. Sonoda, P. Webb, J. D. Baxter, M. S. Skaf and I. Polikarpov, *Biophys. J.*, 2005, **89**, 2011.
74. L. Martinez, P. Webb, I. Polikarpov and M.S. Skaf, *J. Med. Chem.*, 2006, **49**, 23.
75. L. Martinez, I. Polikarpov and M. S. Skaf, *J. Phys. Chem. B*, 2008, **112**, 10741.
76. M. T. Sonoda, L. Martinez, P. Webb, M. S. Skaf and I. Polikarpov, *Mol. Endocrinol.*, 2008, **22**, 1565.

77. J. Shen, W. Li, G. Liu, Y. Tang and H. Jiang, *J. Phys. Chem. B*, 2009, **113**, 10436.
78. L. Celik, J. D. Lund and B. Schiott, *Biochemistry*, 2007, **46**, 1743.
79. K. Ekena, K. E. Weis, J. A. Katzenellenbogen and B. S. Katzenellenbogen, *J. Biol. Chem.*, 1996, **271**, 20053.
80. S. Burendahl, C. Danciulescu and L. Nilsson, *Proteins*, 2009, **77**, 842.
81. M. J. Tsai and B. W. O'Malley, *Annu. Rev. Biochem.*, 1994, **63**, 451.
82. N. Heldring, M. Nilsson, B. Buehrer, E. Treuter and J. A. Gustafsson, *Mol. Cell. Biol.*, 2004, **24**, 3445.
83. E. H. Kong, N. Heldring, J. A. Gustafsson, E. Treuter, R. E. Hubbard and A. C. Pike, *Proc. Natl. Acad. Sci. U. S. A.*, 2005, **102**, 3593.
84. R. L. Rich, L. R. Hoth, K. F. Geoghegan, T. A. Brown, P. K. LeMotte, S. P. Simons, P. Hensley and D. G. Myszka, *Proc. Natl. Acad. Sci. U. S. A.*, 2002, **99**, 8562.
85. M. Peräkylä, *Eur. Biophys. J.*, 2009, **38**, 185.
86. V. Chandra, P. Huang, Y. Hamuro, S. Raghuram, Y. Wang, T. P. Burris and F. Rastinejad, *Nature*, 2008, **456**, 350.
87. E. P. Sablin, A. Woods, I. N. Krylova, P. Hwang, H. A. Ingraham and R. J. Fletterick, *Proc. Natl. Acad. Sci. U. S. A.*, 2008, **105**, 18390.
88. H. Gronemeyer, J. A. Gustafsson and V. Laudet, *Nat. Rev. Drug Discov.*, 2004, **3**, 950.
89. E. Estebanez-Perpina, J. M. Moore, E. Mar, E. Delgado-Rodrigues, P. Nguyen, J. D. Baxter, B. M. Buehrer, P. Webb, R. J. Fletterick and R. K. Guy, *J. Biol. Chem.*, 2005, **280**, 8060.
90. D. J. van de Wijngaart, M. E. van Royen, R. Hersmus, A. C. Pike, A. B. Houtsmuller, G. Jenster, J. Trapman and H. J. Dubbink, *J. Biol. Chem.*, 2006, **281**, 19407.
91. M. N. Zakharov, B. K. Pillai, S. Bhasin, J. Ullor, A. Y. Istomin, C. Guo, A. Godzik, R. Kumar and R. Jasuja, *Mol. Cell. Endocrinol.*, 2011, **341**, 1.
92. M. A. Cunningham, *J. Mol. Model.*, 2012, **18**, 3147.
93. D. J. Mangelsdorf, C. Thummel, M. Beato, P. Herrlich, G. Schutz, K. Umesomo, B. Blumberg, P. Kastner, M. Mark, P. Chambon and R. M. Evans, *Cell*, 1995, **83**, 835.
94. L. Yue, F. Ye, C. Gui, H. Luo, J. Cai, J. Shen, K. Chen, X. Shen and H. Jiang, *Protein Sci.*, 2005, **14**, 812.
95. S. Burendahl and L. Nilsson, *Proteins*, 2012, **80**, 294.
96. M. O. Jensen, V. Jogini, D. W. Borhani, A. E. Leffler, R. O. Dror and D. E. Shaw, *Science*, 2012, **336**, 229.
97. G. R. Bowman, V. A. Voelz and V. S. Pande, *J. Am. Chem. Soc.*, 2011, **133**, 664.

CHAPTER 5

Docking, Screening and Selectivity Prediction for Small-molecule Nuclear Receptor Modulators

RUBEN ABAGYAN*, WINSTON CHEN AND
IRINA KUFAREVA

Skaggs School of Pharmacy and Pharmaceutical Sciences, University of
California, San Diego, 9500 Gilman Drive, La Jolla, CA 92093, USA
*E-mail: rabagyan@ucsd.edu

5.1 Introduction

Nuclear receptors (NRs) represent a family of ligand-dependent transcription factors, united by common functional mechanism and domain architecture. At least 13% of marketed drugs target NRs.[1] The two key domains, the ligand-binding domain (LBD) and the DNA-binding domain (DBD), are conserved in most NRs and are surrounded by the more variable regions (N-terminal, hinge region and C-terminal domains). NR LBDs recognize, sometimes with picomolar affinity, small membrane-diffusible endogenous hormones, metabolites or xenobiotics. A ligand binding event induces reorganization of complex intra- and intermolecular interactions, which ultimately leads to the formation of a complex between the NR DBD and the hormone response element of the target DNA and consequently to either initiation or repression of the target gene transcription. Nuclear receptors are found in metazoans

RSC Drug Discovery Series No. 30
Computational Approaches to Nuclear Receptors
Edited by Pietro Cozzini and Glen E. Kellogg
© The Royal Society of Chemistry 2012
Published by the Royal Society of Chemistry, www.rsc.org

(animals), including fish, birds and mammals. Humans have 48 NRs (Figure 5.1) and numerous splice variants.

While most human NRs (with the notable exception of pregnane X receptor) are fairly specific, many of the marketed NR modulators exhibit varying degrees of polypharmacology. Furthermore, it has been demonstrated that some environmental substances and their metabolites bind to NRs, resulting in the disruption of endocrine signaling pathways and developmental defects.[2,3] Understanding the structural and molecular basis of the specific binding of small molecules to the LBDs of NRs represents an important challenge for both structural computational chemistry and green chemistry. Translation of this understanding into predictive chemistry-unbiased and training-independent models may lead to a new generation of safer therapeutics and early alerts concerning dangerous environmental chemicals.

Therapeutic agents and environmental chemicals may affect NR-mediated gene expression in several ways, *e.g.*, by mimicking endogenous hormones (full and partial agonists), sterically blocking activation by endogenous hormones (antagonists) or inhibiting the basal ligand-independent transcription activation (inverse agonists). While any of these activities requires binding to the LBD, the specific mode of action of a chemical modulator (agonism or antagonism) may depend on many factors. Some modulators block binding and activation of the receptor by an endogenous agonist when the latter is

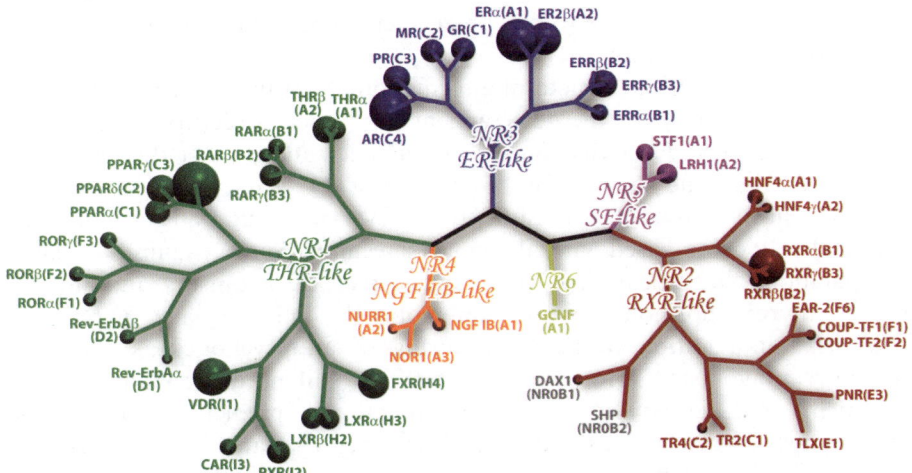

Figure 5.1 The phylogenetic tree for the 48 human nuclear receptors based on sequence similarity of their LBDs. The receptors are classified into seven subfamilies (NR0–NR6), which are further divided into groups. The nodes on the tree are marked by the abbreviated conventional name of the receptor (*e.g.*, PPARγ for peroxisome proliferator-activated receptor γ) and also their group and member ID within the subfamily (*e.g.*, NR1C3). The number of LBD structures available in the PDB for each receptor at the time of publication is given in Table 5.1 and is represented by the volume of the spheres.

present in sufficient quantities, therefore acting as competitive antagonists. However, in the absence of the endogenous agonist, the same compound can weakly activate the receptor, acting as a partial agonist. The effects of the drug may also depend on the fine balance of receptor isoforms, co-activators and co-repressors, number and location of hormone response elements, and the presence and activation state of other transcription factors in a specific tissue or a cell type.[4] These factors explain the phenomenon of tissue selectivity and enable the development of tissue-selective NR modulators. Structurally, such compounds are believed to affect the equilibrium between active and inactive conformations of their target receptors, which in turn is interpreted by the organism in a tissue-specific manner. Tissue-selective modulators are known for several receptors, including estrogen receptor (ER),[5] androgen receptor (AR),[6] glucocorticoid receptor (GR)[4,7] and progesterone receptor (PR).[8]

NRs evolved to interact with different classes of molecules (*e.g.*, steroid hormones, fatty acids or phospholipids); therefore, they manifest substantial sequence and structural variability in the LBD; however, within the families or subfamilies of related receptors, ligand–receptor cross-reactivity frequently occurs. Some of this cross-reactivity results in physiological or adverse effects of therapeutics. The concept of NR subtype cross-reactivity is illustrated by the androgenic effects of early contraceptives.[9,10] Even after the extensive subtype-selective optimization that these drugs underwent in recent years, the beneficial side effects of combined oral contraceptive pills, and also unpleasant withdrawal symptoms, are partially due to off-target interaction of progestogens with the mineralocorticoid receptor.

The growing wealth of structural information about the NR LBDs provides a unique opportunity for the development of physical structural models that explain the observed activity and cross-reactivity and predict the endocrine effects of drugs and xenobiotics. For the last 15 years, individual crystal structures have been used successfully to discover new drug leads or molecular probes by docking-based virtual ligand screening. Some examples include computational identification of retinoic acid receptor agonists[11] and antagonists,[12] thyroid hormone receptor antagonists[13] and pregnane X receptor agonists,[14] in addition to the repurposing of antipsychotics as androgen receptor antagonists.[15] However, methods are still needed to improve the specificity and selectivity of the computational predictions, to profile compounds against the extended nuclear receptor panel *in silico* and to predict functional activity.

This chapter summarizes and analyzes the structures of the LBDs of the human NRs, and also the two-dimensional (2D) and three-dimensional (3D) chemical information about their chemical modulators. We demonstrate that multiple co-crystal structures of the same LBDs may improve the ability of the ligand docking and scoring methods[16] to recognize chemically diverse modulators and may result in more accurate chemistry-unbiased predictive models of activity. We also show that the 3D coordinates of co-crystallized modulators alone can be converted into 3D pharmacophore property models[17] that, in turn, can be used to recognize potential modulators by docking/ scoring. These approaches can also be treated as complementary. The aggregated data, models and methods may lead to the discovery of drugs with improved NR profiles or the early discovery of endocrine risks of environmental chemicals.

5.2 Chemicals Targeting Nuclear Receptors

5.2.1 Chemical Specificity: from Highly Specific to Highly Promiscuous

The ligand-binding pockets in most NRs are tuned to fit specific endogenous chemicals. Endogenous activators have been described for more than 30 receptors. In agreement with the size distribution and chemical properties of the NR ligand-binding pockets, the endogenous activators are usually relatively small (20–30 heavy atoms) and lipophilic (log$P$ between 1 and 7) molecules. They cover a large and diverse chemical space and include lipids, fatty acids, retinoids, steroid hormones, prostaglandins, *etc.* Although in the comprehensive 2006 overview of NR nomenclature,[18] 21 of the 48 receptors were described as orphan, by now at least one sub-micromolar endogenous or exogenous activator has been identified for 10 of these orphan receptors, which is reflected in either crystallographic or medicinal chemistry databases (Table 5.1). Several receptors that have only been crystallized in the apo form manifested the absence of the binding pocket, which may be, but not always is, the inherent feature of the receptor relating to its inability to bind any ligands. Nerve growth factor IB (NGF IB, also known as NUR/77 or NR4A1) is an example of a receptor whose X-ray crystal structure does not have any internal cavity; however, it is now known that cytosporone B is its naturally occurring agonist[19] and additional synthetic NGF IB agonists have been also identified.[20] The apparent inconsistency is probably due to the natural flexibility of the NGF IB binding pocket that tends to collapse in a crystallographic environment in the absence of the ligands. NGF IB provides a clear example of the conformational plasticity of the NR LBDs and the widely varying utility of crystallographic structures for augmenting our understanding of NR ligand binding and specificity.

Depending on the degree of the conformational plasticity of the LBD, and also on the precise arrangement of the pharmacophore features of the residues in the pockets, some nuclear receptors may be more picky towards their ligands than others. The distribution of ligand MQN properties (shape, size and atom composition[21]) varies from extremely wide in pregnane X receptor (PXR) to very narrow in vitamin D receptor (VDR) (Table 5.2). This kind of ligand property distribution analysis not only reflects the biological properties of the receptors (*e.g.*, it characterizes PXR as a promiscuously activated transcription factor), but, as we will show below, also gives preliminary insights into the level of difficulty of ligand activity prediction for each individual receptor.

Table 5.1 Structural and bioactivity data (available in public databases as of December 2011) for the 48 human nuclear receptors. Of the 39 receptors with X-ray structures of the LBD in PDB, most are represented by (i) multiple crystal structures, thus providing important information about pocket flexibility, and (ii) multiple co-crystallized ligands, thus allowing the derivation of the mean property fields. Additional ligand activity information can be derived from chemical databases, e.g., ChEMBL. Receptors for which no ligands have been identified to date are italicized.

Protein name	Nomenclature	Abbreviation	UniProt	PDB No.[a]	No. of seeds	No. of ChEMBL actives: $-log_{10}(activity)$				
						>6	>7	>8	>9	>10
Nuclear receptor DAX-1	*NR0B1*	*DAX1*	*NR0B1*	1(m)						
Orphan nuclear receptor SHP (small heterodimer partner)	NR0B2	SHP	NR0B2			1				
Thyroid hormone receptor α	NR1A1	THRα	THA	6	4	145	74	33	12	2
Thyroid hormone receptor β	NR1A2	THRβ	THB	16	9	231	157	96	53	10
Retinoic acid receptor α	NR1B1	RARα	RARA	4	4	115	68	27	11	7
Retinoic acid receptor β	NR1B2	RARβ	RARB	2	2	150	103	42	14	5
Retinoic acid receptor γ	NR1B3	RARγ	RARG	9	8	153	100	38	11	6
Peroxisome proliferator-activated receptor α	NR1C1	PPARα	PPARA	13	13	623	269	76	35	34
Peroxisome proliferator-activated receptor δ	NR1C2	PPARδ	PPARD	21	18	456	307	125	4	
Peroxisome proliferator-activated receptor γ	NR1C3	PPARγ	PPARG	80	68	902	478	198	46	34
Rev-erbA α (EAR-1)	*NR1D1*	*Rev-ErbAα*	*NR1D1*	1						
Rev-erbA β (EAR-1R)	NR1D2	Rev-ErbAβ	NR1D2	3	2					
Retinoid-related orphan receptor α	NR1F1	RORα	RORA	2	2					
Retinoid-related orphan receptor β	NR1F2	RORβ	RORB	3 (r)	3					
Retinoid-related orphan receptor γ	NR1F3	RORγ	RORG	4	4					
Oxysterols receptor LXR-β	NR1H2	LXRβ	NR1H2	8	7	352	233	90		
Oxysterols receptor LXR-α	NR1H3	LXRα	NR1H3	7	7	381	204	76	3	
Bile acid receptor (farnesoid X-activated receptor)	NR1H4	FXR	NR1H4	24	24	213	86	11		
Vitamin D₃ receptor	NR1I1	VDR	VDR	46 (hr)	41	76	65	41	31	14
Orphan nuclear receptor PXR (pregnane X receptor)	NR1I2	PXR	NR1I2	10	7	19	9	4		
Constitutive androstane receptor	NR1I3	CAR	NR1I3	4 (hm)	4	2				
Hepatocyte nuclear factor 4 α	NR2A1	HNF4α	HNF4A	3	2					

Table 5.1 (*Continued*)

Protein name	Nomenclature	Abbreviation	UniProt	PDB No.[a]	No. of seeds	No. of ChEMBL actives: $-\log_{10}(activity)$				
						>6	>7	>8	>9	>10
Hepatocyte nuclear factor 4 γ	NR2A2	HNF4γ	HNF4G	1	1					
Retinoid X receptor α	NR2B1	RXRα	RXRA	39	22	263	168	77	7	3
Retinoid X receptor β	NR2B2	RXRβ	RXRB	2	2	93	63	23	1	1
Retinoid X receptor γ	NR2B3	RXRγ	RXRG	1		87	58	11		
Testicular receptor 2	NR2C1	TR2	NR2C1							
Testicular receptor 4	NR2C2	TR4	NR2C2							
Nuclear receptor TLX (protein tailless homolog)	NR2E1	TLX	NR2E1	1						
Photoreceptor-specific nuclear receptor	NR2E3	PNR	NR2E3			15				
COUP transcription factor 1 (EAR-3)	NR2F1	COUP-TF1	COT1							
COUP transcription factor 2 (apolipoprotein A-I regulatory protein 1)	NR2F2	COUP-TF2	COT2	1						
V-erbA-related protein 2 (EAR-2)	NR2F6	EAR-2	NR2F6							
Estrogen receptor α	NR3A1	ERα	ESR1	67	54	1081	691	374	111	7
Estrogen receptor β	NR3A2	ERβ	ESR2	31	28	1057	698	226	43	4
Estrogen-related receptor α	NR3B1	ERRα	ERR1	4	2	30	17	6		
Estrogen-related receptor β	NR3B2	ERRβ	ERR2			16	4	1		
Estrogen-related receptor γ	NR3B3	ERRγ	ERR3	18	8	10	4	1		
Glucocorticoid receptor	NR3C1	GR	GCR	12	7	1095	870	492	77	1
Mineralocorticoid receptor	NR3C2	MR	MCR	11	6	146	34	10	2	
Progesterone receptor	NR3C3	PR	PRGR	16	14	794	534	200	62	2
Androgen receptor	NR3C4	AR	ANDR	64	23	825	535	211	35	2
Nuclear hormone receptor NUR/77 (testicular receptor 3)	NR4A1	NGF IB	NR4A1	2		7	1	1		
Orphan nuclear receptor NURR1	NR4A2	NURR1	NR4A2	1						
Neuron-derived orphan receptor 1	NR4A3	NOR1	NR4A3							
Steroidogenic factor 1	NR5A1	STF1	STF1	5	4	28	6			
Liver receptor homolog 1	NR5A2	LRH1	NR5A2	7 (hm)	4	14	7			
Germ cell nuclear factor	NR6A1	GCNF	NR6A1							

[a]Structures of human (h), mouse (m) and rat (r) receptors listed.

Table 5.2 Ligand promiscuity indices calculated as normalized pairwise MQN distances within the ChEMBL activity datasets for each receptor. Receptors are listed in the order of decreasing promiscuity.

Receptor	Ligand promiscuity index
PXR	1.006
PR	0.391
RARα	0.385
MR	0.373
ERRβ	0.362
FXR	0.358
GR	0.353
AR	0.340
RXRβ	0.338
RARβ	0.328
RARγ	0.326
RXRγ	0.323
ERRα	0.310
LXRβ	0.299
THRα	0.285
THRβ	0.285
ERβ	0.283
ERα	0.283
LXRα	0.282
PPARα	0.246
ERRγ	0.239
RXRα	0.237
PPARδ	0.218
PPARγ	0.207
VDR	0.139
NGF IB	0.000[a]
STF1	0.000[a]

[a]Insufficient ChEMBL activity data.

5.2.2 Agonists, Antagonists, SNuRMs

Most endogenous chemicals targeting NRs function as agonists and lead, depending on the receptor type, to either activation or repression of target gene transcription. In contrast, therapeutic agents belong to several classes: whereas some mimic the action of the endogenous agonists (*e.g.*, rosiglitazone and bexarotene), others bind to the target domain, preventing downstream signaling or transcriptional activity changes. Moreover, they prevent stimulation of the receptor by the endogenous agonists by competing for the same binding site (*e.g.*, mifepristone, RU486). Drugs with a mixed agonist–antagonist profile often modulate the signaling by their target receptor in a tissue-selective manner and are called selective nuclear receptor modulators (SNuRMs). Following the serendipitous discovery of tamoxifen, as a tissue-selective estrogen receptor modulator that acts as agonist in the bone and as an antagonist in breast,[5,22] with both actions leading to beneficial therapeutic effects, there has been growing interest in the development of SNuRMs for other receptors.

Although the early therapeutic agents targeting NRs were close chemical analogs of the endogenous hormones, chemically divergent modulators are now known for best studied receptors, *e.g.*, non-steroidal agonists and antagonists of the steroid receptors. In many cases, drugs of novel chemical classes are characterized by improved pharmacokinetic properties compared with the analogs of endogenous agonists.

The chemical features that distinguish agonists from antagonists are not always obvious, in agreement with the complex structural basis for LBD activation (see Section 5.3.2). Estrogen receptor antagonists are distinguished from agonists by a bulky ionizable 'side chain' that, in the bound state, extends towards the opening of the ER binding pocket, sterically interfering with the active conformation of helix 12. However, a large number of NR antagonists do not have this bulky side chain (*e.g.*, flutamide). Moreover, even with closely related receptors, a single chemical may act as an agonist for one and an antagonist for the other. For example, progesterone is an agonist for the progesterone receptor but an antagonist for the closely homologous mineralocorticoid receptor, while the synthetic estrogen THC acts as an agonist and an antagonist for the estrogen receptor subtypes α and β, respectively.[23]

5.2.3 Analysis of Cross-reactivity

A comparison of chemical activity sets for individual NRs also provides insight into the degree of their cross-reactivity, *i.e.*, the ability of different NRs to be activated by the same ligands. Figure 5.2a illustrates the principles of calculation of cross-reactivity potential of pairs of NRs based on similarity of their ChEMBL ligand activity sets in the spirit of ref. 24. Given two sets of chemical activities measured against receptors A and B, a pairwise measure of compound similarity (in the presented plot, Tanimoto distance on chemical fingerprints) is calculated for all compound pairs within the activity set A, all pairs within the set B and all pairs between the two sets. Three cumulative distribution curves are constructed, each plotting the fraction of pairs in the set (or between the sets) that fall under a specified distance cutoff against that cutoff. Next, the distribution of pairwise distances between the two sets is compared with each of the intra-set distributions. The two sets appear well separated in the chemical space if at low chemical distance cutoffs, there are significantly fewer inter-set pairs below the cutoff than there are intra-set pairs. In this scheme, the definition of the 'low chemical distance cutoffs' is rather arbitrary; we chose to introduce a continuous weight function $W(D) = \exp(-200D^4)$ that gives a weight of ~ 1 to chemical distances $0 \leq D < 0.2$, quickly degrades to zero for $0.2 \leq D < 0.4$ and is equal to zero for $D > 0.4$.

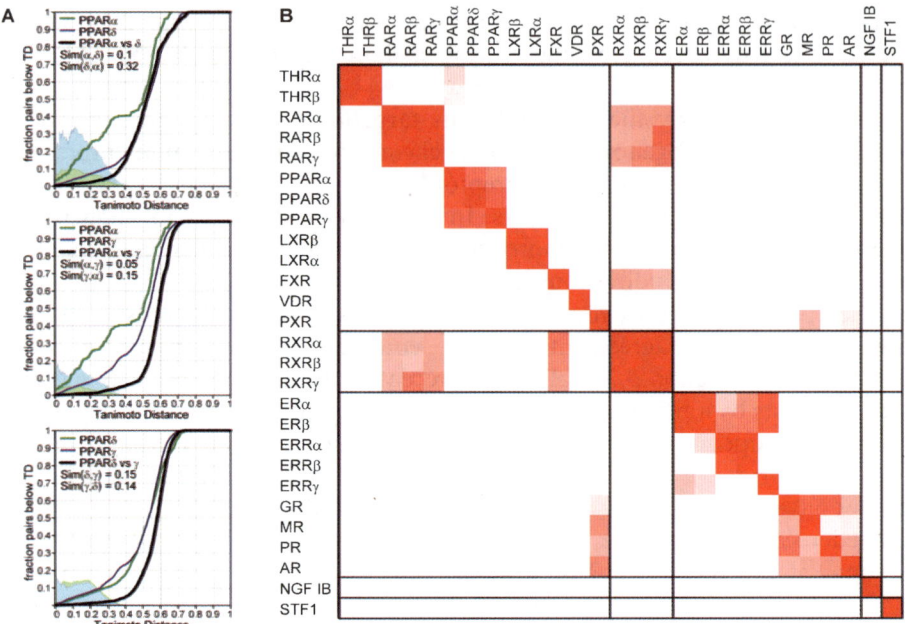

Figure 5.2   Chemical cross-reactivity potential of the human nuclear receptors. (a) Fingerprint similarity of experimental ligand sets of two receptors can be used to characterize quantitatively their common pharmacology in the spirit of ref. 24 (see details in text). (b) Matrix of pairwise pharmacological similarities for 27 human receptors with available ChEMBL actives. As expected, closely related receptor subtypes often share common pharmacology; however, a substantial overlap is also observed between more distant receptor pairs.

The proposed chemical set similarity measure takes into account not only the number of similar compounds between the sets, but also the degree of 'chemical spread' within each of the sets. Comparing a 'tighter' set with a set with wider 'spread' results in a larger value than comparing the same two sets in reverse order. This is because a related compound has a higher probability of accidentally hitting a target that is inherently more promiscuous. The concept is illustrated by a pairwise comparison of the chemical activity sets for the three peroxisome proliferator-activated receptors (PPARs), α, δ and γ. This comparison demonstrates that there is a high probability of hitting the δ isoform with a compound from the α set [Sim(δ,α) = 0.32] and a lower but still significant probability of hitting the α isoform with a compound from the δ set [Sim(α,δ) = 0.1]. This correlates with the well-documented difficulties in the design of α- and δ-selective PPAR agonists, and also with smaller number of compounds that are selective for PPARα versus PPARδ.

At the time of writing this chapter, low nanomolar modulators could be found in ChEMBL for 27 human nuclear receptors. The 27 × 27 matrix of ChEMBL set similarities (Figure 5.2b) calculated as described above reveals non-trivial relations in addition to the expected cross-reactivity between the closely related receptor subtypes. For example, it shows that the ligands for the NR3C family of receptors (GR, MR, PR and AR) have a substantial probability of hitting the pregnane X receptor.

We show in this chapter that computational models based on 3D crystallographic structures of the NR LBDs are highly instrumental not only for the prediction of ligand activities, but also for ligand profiling and for identification of subtype-selective modulators.

5.3 Nuclear Receptor Ligand-binding Domains

5.3.1 Structural Coverage

As of December 2011, X-ray structures had been solved for 37 of the 48 human NRs, often in complexes with multiple high-affinity modulators. With only a few exceptions, each receptor is represented by multiple (co-)crystal structures. Even for the single-entry receptors, one can typically find multiple instances of the protein in an asymmetric unit (*e.g.*, four variations in the PDB entry 2GL8 for RXRγ). Each of ERα, VDR, AR, RXRα and PPARγ are represented by over 30 PDB entries with many more individual variations of the pocket. This increases our understanding of possible induced fit effects, and also, in some cases, the structural mechanisms underlying the phenomena of agonism, antagonism or allosteric modulation. The wealth of information about ligand-dependent structural variability of the protein binding pockets and the resulting functional response diversity of the proteins is well represented in the Pocketome database,[25] a large part of which is composed of the NR LBDs.

All LBDs of the NRs share a similar topology, shown in Figures 5.3 and 5.4. The LBDs crystallized so far reveal a highly structurally conserved α-helical fold with a deep well-protected 'abdominal' cavity; the 'entry door' to the cavity is formed by helices H11–H12 and the 'back' wall is an anti-parallel β-sheet of variable size.

5.3.2 Structural Basis of Agonism and Antagonism

NR agonists usually act by stabilizing a specific conformation of the LBD. This conformation is characterized by a snug fit between a hydrophobic residue at the base of H12 and four other residues in helices H3, H5 and H11. The obtained 'pyramid' of interactions (Figure 5.5) effectively leads to the formation of the co-activator peptide-binding interface on the surface of helices H3, H4 and H12 and consequent co-activator recruitment. In contrast, small molecules that prevent the formation of the pyramid act as either pure antagonists or tissue-selective modulators, depending on the degree of pyramid destabilization. The latter class of molecules is the most interesting: they only partially destabilize the active 'pyramid' conformation so that it remains one of several equilibrium conformations of the LBD, and the downstream functional

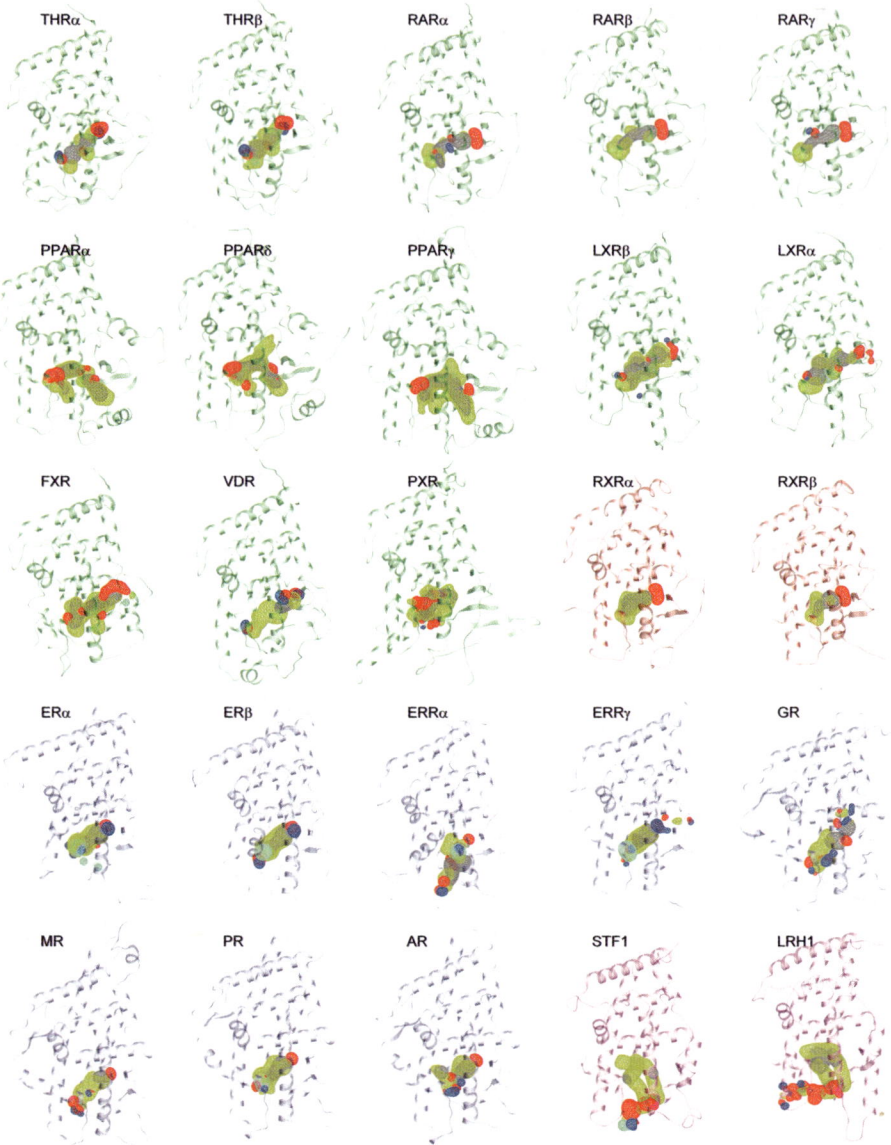

Figure 5.3 The 25 human NRs for which all three of structures, co-crystal (seed) ligands and ChEMBL activity sets are available.

effects become entirely dependent on the relative concentrations of co-activators, co-repressors and other binding partners in the specific tissue.

Destabilization of the 'pyramid' by drugs may have a more or less pronounced effect on the overall conformation of the LBD. In the NR3A and NR3B family of receptors (estrogen and estrogen-related receptors), antago-

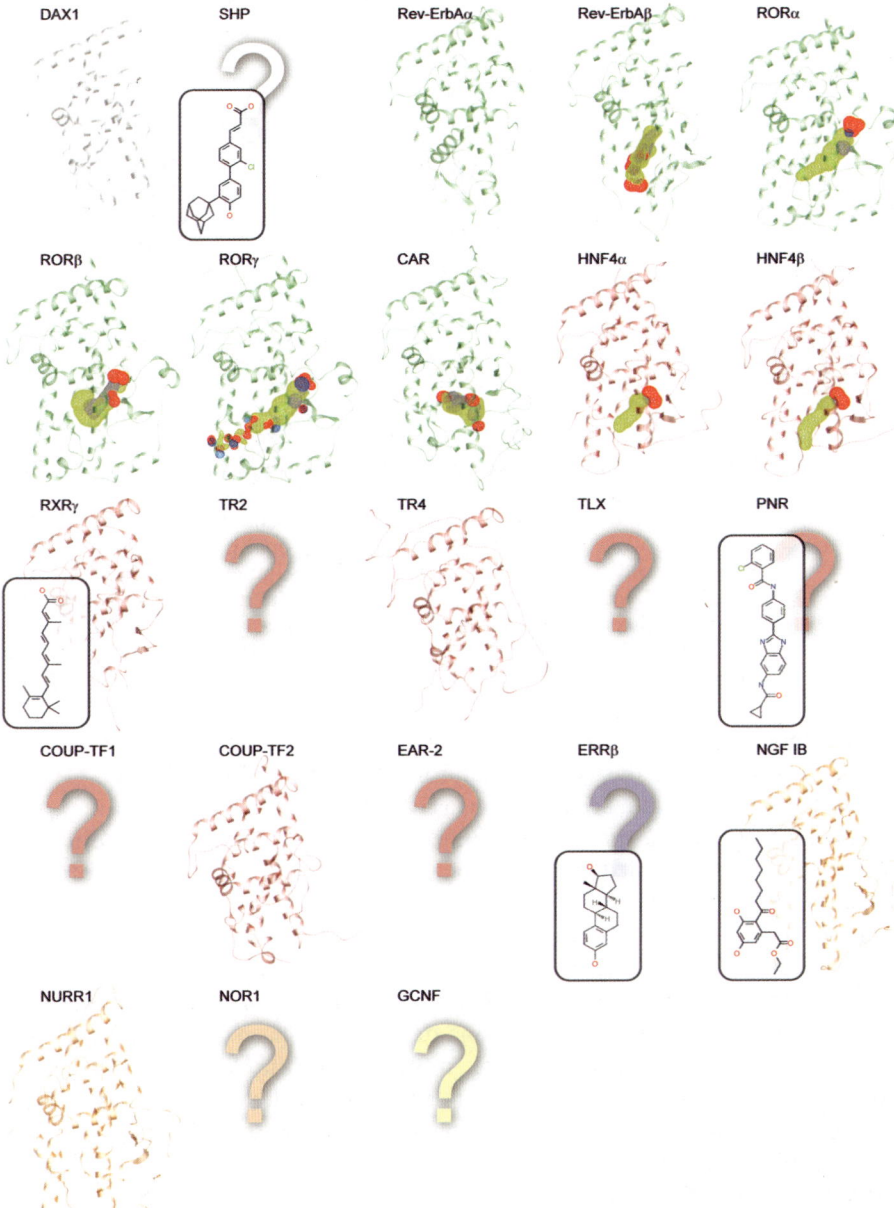

Figure 5.4 The 23 human NRs lacking information for 3D activity model generation. Receptors with no available structures are shown with question marks. At least one endogenous or synthetic agonist is known for the seven receptors shown as structures with APF densities and the five receptors shown with a representative chemical structure. It is not clear whether the activity of the remaining 11 receptors is ligand dependent.

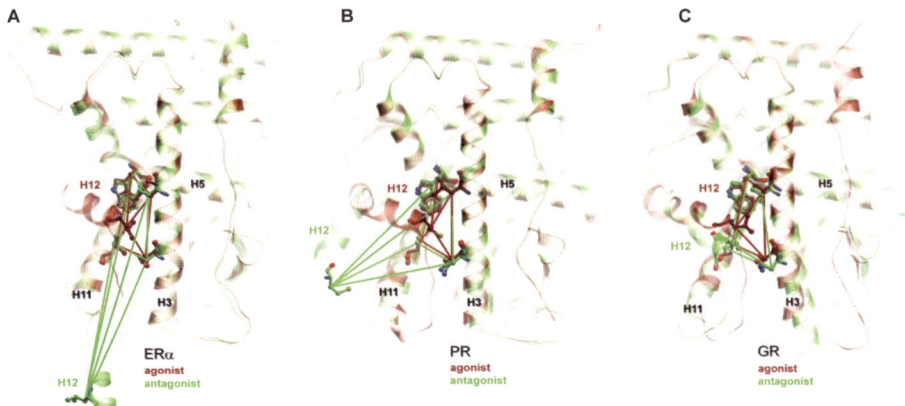

Figure 5.5   Structural basis of agonism and antagonism in NRs. Agonists stabilize
the LBD in an active conformation (dark red) characterized by a
conserved network of interactions between residues in helices 3, 5, 11 and
12. The network can be visualized as a tetragonal pyramid where H12 is
the vertex and structurally conserved residues in helices 3, 5 and 11 form
the base. Molecules that destabilize the 'pyramid' to different degrees act
as antagonists, sometimes in a tissue-selective manner. (a) Estrogen
receptor α (ERα); (b) progesterone receptor (PR); (c) glucocorticoid
receptor (GR).

nist binding often leads to the detachment and a large (∼30 Å) swing
movement of H12. In the NR3C receptors (GR, MR, PR and AR), H12 is
followed by a β-strand that keeps H12 attached to the domain core;
consequently, the conformational changes brought by antagonist binding are
modest and manifest as smaller displacements of H11 and the base of H12
(Figure 5.5).

Mutations of 'pyramid' residues may result in a receptor that is either
constitutively active or accommodates wild-type receptor antagonists, while
retaining the productive co-activator-binding conformation. These mutations
have a selective advantage in hormone-dependent cancers: for example, the
well-characterized T877A and W741L mutations in the AR that evolve in a
prostate cancer patient as a result of treatment with anti-androgens (*e.g.*,
flutamide, bicalutamide) and confer agonist activity on these drugs, belong to
the class of pyramid residue mutations.

In many cases, the effect of destabilization of the active conformation by
antagonists or SNuRMs is so subtle that it can be completely or partially reversed
by the crystallographic environment of the LBD in an X-ray structure.
Consequently, in crystallography, antagonists and SNuRMs are sometimes
observed bound to the active conformation of the target domain.[26] In the context
of structure-based prediction of ligand activity, this means that apart from the
ligand itself, an LBD X-ray structure may not necessarily carry the information
about the functional class of its cognate modulator. Pocket structures crystallized

with antagonists usually recognize agonists well in virtual ligand screening. The opposite may also be true, although it occurs less frequently.

5.4 Computational Prediction of Nuclear Receptor Modulators

While crystallography of the NR LBDs assists our understanding of the structural principles of ligand interactions with their target receptors, the ultimate goal of computational NR biology is the prediction of affinity, activity and selectivity profiles for novel compounds. For that, the crystal structures need to be converted into predictive models and carefully benchmarked in retrospective applications. The two complementary approaches to the problem address it from the side of the binding pocket and from the side of the co-crystallized ligand. In both cases, the prediction is based on the docking of flexible compounds, with consequent scoring of the lowest energy compound poses[27] and ranking of the compounds based on the scores obtained. Below we outline these approaches, the applicability domains of the resulting models and difficulties associated with model generation.

5.4.1 Improving the Docking and Scoring Accuracy to Ligand-binding Domains

Flexible ligand docking into NR models can be categorized into the following cases of progressively more uncertain and error-prone procedures:

(a) (the safest) docking to multiple, conformationally distinct models derived from multiple co-crystal structures of the same protein;
(b) docking to a single model derived from one crystal structure and compound screening;
(c) docking to homology models;
(d) docking of an antagonist to a LBD model derived from co-crystal structure with agonist;
(e) docking to allosteric sites.

Each category has distinct challenges and the expected success rate in terms of both ability to predict the correct binding pose and use the docking/scoring for predicting activity of new compounds deteriorates very quickly from almost perfect results for the majority of nuclear receptors in the (a) category to a risky gamble for (c), (d) and (e).

5.4.1.1 Docking to Pockets Obtained from Multiple Co-crystal Structures

From observing the multiple co-crystal structures of a single LBD, it becomes obvious that a certain conformational variability of the pocket is an inevitable attribute of the induced fit of ligand binding. In extreme cases (*e.g.*, NGF IB), the crystal structure pocket can collapse entirely and prohibit any ligand binding, despite the fact that the receptor is known to bind and be activated by ligands (Section 5.2.1). In more subtle cases, it often appears that some pockets favorably dock and score multiple ligands that belong to different chemical classes, others only recognize a single chemical and down-score the other high-affinity binders, and the rest are not selective at all in scoring the known binders against random inactive molecules of matching size and physico-chemical properties. In other words, the screening performance of individual X-ray structures of a single LBD may vary greatly due to major or minor variations in the pocket conformation.

In the situation when a sufficiently large amount of medicinal chemistry data is available for the given receptor, in the form of chemicals and their affinities and activities, the quality and utility of a pocket model can be evaluated by retrospective screening. In this procedure, the known actives are mixed in a single dataset with decoy molecules of matching size, molecular weight, atom counts, log*P*, *etc.* (the so-called *property-matched decoys*), followed by docking and scoring to the model. A model with good selectivity consistently ranks known actives better than decoys, resulting in ideal (or close to ideal) shape of the receiver operating characteristic (ROC) curve with the area under the curve (AUC) being close to 1. In contrast, a bad model ranks actives and decoys equally well, resulting in an approximately straight diagonal ROC curve (AUC = 0.5) or sometimes even ranks actives worse than decoys (AUC close to zero).

Because the conformational variants of a single LBD may recognize distinct classes of chemicals, it is important to include a representative ensemble of those essential conformations as multiple independent models (so-called ensemble docking, *e.g.*, ref. 28), as concurrently present 'four-dimensional' grid potentials.[29]

We recently tested 17 NRs for single *versus* multiple receptor performance and the multiple conformation performance was consistently better than an average single conformation.[16] However, it is not as simple as 'the more the better'. Three important non-trivial caveats need to be stated.

- First, as described above, for receptors with multiple experimental crystal structures, the screening performance measured as ability to separate actives and decoys by docking and scoring may be highly variable between individual models.[16] The choice of a structure representing the pocket can dramatically change the screening outcome.
- Second, including *all* models in the ensemble and selecting the lowest score usually decrease the success rate for screening and recognition (Figure 5.6), because each additional crystal structure and conformation not only gives a possible induced conformation, but also increases the chance for a decoy to have lower score, therefore increasing the 'noise'. Since the number of decoys (or non-actives) far exceeds the number of actives, this factor becomes dominant in a screen.

- Third, partially because of the above reason, it is possible that a single conformation gives a better ROC AUC value than multiple structures with a given set of actives and decoys. For example, a single refined model for PPARγ produces impressive AUCs of 0.94 and 0.92 for the DUD[30] and Wombat[31] sets, respectively.[32]

Practically, we employ the following recipe:

- Choose a set of actives (A) and decoys (D) for a given pocket from an activity database, *e.g.*, ChEMBL.[33]
- Perform initial refinement (hydrogen placement, Asn, Gln, His rotations) of each model with a cognate ligand and remove ligands to create a pocket.[34]
- Evaluate the ICM docking and scoring performance of each available co-crystal structure, by retrospective screening of known actives against relevant decoys and measuring the resulting area under ROC curve (AUC).
- Choose one or several models that collectively provide the best separation between actives and decoys. The aggregated score is the lowest score for a given compound.

As can be seen from the distributions in Figure 5.6, this approach provides excellent models for over 60% of the nuclear receptors. These models assign docking scores better than the 0.1% of the decoys for anywhere from 20 to 99% of the ChEMBL-annotated actives even with over 20-fold more decoys with matching properties.

However, for certain receptors, such as the glucocorticoid, mineralocorticoid, pregnane X and liver X receptors, these selected and lightly refined multi-conformational models are not sufficient. Some of the difficulties are clearly related to the larger cavities, naturally designed to accommodate a wider variety of chemicals. In other cases, potential reasons for poor screening performance could include the insufficient refinement of the cognate structures or insufficient conformational repertoire of pockets. Finally, avenues for screening performance improvement also include separation of the distinct functional states within the ensembles, accounting for essential water molecules and their displacement and improving the accuracy of 2D to 3D conversion of complex steroid cores.

Each ROC curve shows how actives and decoys are separated by the predicted docking scores. However, the range and the distribution of those scores depend on a protein and, to a lesser extent, on a model representing the protein. To predict an activity of a new compound, one needs to know if the score of the compound fits within the decoy score distribution or if it is significantly better than decoy scores, which justifies any assumptions made about its activity. In order to provide a compact and robust representation of the decoys score distribution, one may numerically fit it into an analytical function and store the resulting fit parameters. The parameters can then be used to calculate both the normalized score for a new compound or a probability of its inactivity.

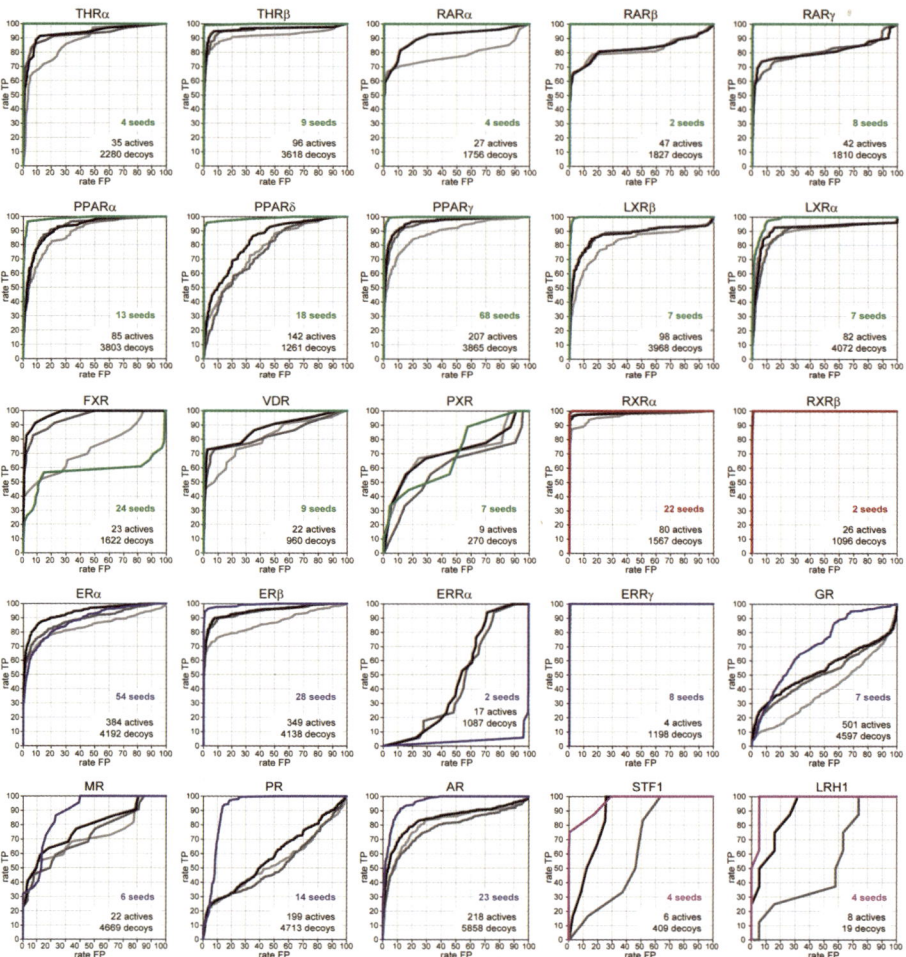

Figure 5.6   Virtual screening performance of the NR models based on optimal pocket conformation ensembles (black curves), all-inclusive ensembles (dark-grey curves), best single conformation (light-grey curves), and ligand atomic property fields (APF, colored curves). The actives were selected from ChEMBL activity datasets with an adaptive activity cutoff: 10 nM for receptors with more than 20 low-nanomolar modulators and decreasing to at most 100 nM for receptors with fewer high-affinity modulators. The decoys were selected from virtual databases of commercially available compounds based on the similarity of chemical properties (molecular weight, log*P*, atom counts, charge, *etc.*). The number of actives and decoys is shown on each plot. For APF models, the number of seed molecules used to build each model is also shown. ROC curves provide retrospective evaluation of the screening performance of each model in question.

In summary, the refined multi-conformational models represent an ultimate recognition device for new chemistry once one or several co-crystal structures have been determined. They can be used for virtual ligand screening, activity prediction and individual compound profiling, once the parameters of the noise distribution have been derived from a large number of decoy scores.

5.4.1.2 Docking and Screening Against a Single Crystal Structure

Docking molecules into a single crystal structure was traditionally viewed as the mainstream docking and screening tool. However, there are three important caveats to be taken into consideration.

First, a model deposited in the PDB may need additional refinement or modeling within the limits of the experimental electron density to address five issues: (i) placement and optimization of rotatable hydrogens; (ii) optimization of orientations of Asn, Gln and His and tautomeric/protonation states of His (six states of each His side chain are compatible with the electron density); (iii) optimization of the parts of the model in the regions of unclear electron density; (iv) overall small positional adjustments to ensure more realistic packing with the co-crystallized ligand; and (v) selection of water molecules to be retained in the pocket. All these changes do not change the fit of the model to the electron density but produce a much more realistic and, in the end, efficient model for screening.

Second, although the optimized model may now be used for virtual ligand screening of new chemicals that may potentially bind to this pocket, it is relatively useless in predicting the activity of a new chemical scaffold, since a single receptor conformation is expected to recognize only a subset of the actives and have a sizeable fraction of false negatives.

Third, the models may be further manipulated to accommodate larger antagonists by moving parts of the H11–H12 loop and helix 12.

The internal coordinate (ICM) docking to a single conformation has been used successfully for screening for new chemical leads over the last 12 years. In 2000, two new retinoic acid receptor antagonists were discovered by virtual ligand screening of a model in which helix 12 was moved away to resemble the antagonist-bound conformations of estrogen receptor.[12] In 2001, 150 000 molecules were screened against a single RAR model without modification, 30 molecules were tested and two new potent agonists ($K_i < 50$ nM) were discovered.[11] In 2003, a homology model was built for the thyroid hormone receptor LBD and it was used to screen a 250 000 compound library. Out of 64 compounds tested, 14 were found with antagonistic activity ranging from 1.5 to 30 µM.[13]

As we depart from a refined model into predicted states and homologous proteins, the success rate is no longer guaranteed. It may also require some 'ligand guidance' to make sure that the model at least binds several reference molecules favorably. In the first use of this technique,[35] two antagonist-bound models of the androgen receptor were generated using the ligand-guided procedure and the resulting models were used to identify androgenic activities of some antipsychotic drugs. This is also an example of docking-based drug repurposing.

In conclusion, although virtually screening a compound database against a single crystal structure can be a possible screening tool with over 5–10% early success rate, this screen is bound to miss many active molecules. Furthermore, the models may need refinement and sometimes further manipulation to accommodate alternative functional states.

5.4.2 Ligand-derived Atom Property Fields from Co-crystal Structures as Activity Predictors

For 25 NRs, namely THR α and β, RAR α, β and γ, PPAR α, δ and γ, LXR α and β, FXR, VDR, PXR, RXR α and β, ER α and β, ERR α and γ, GR, MR, PR, AR, STF1 and LRH1 (Table 5.1), more than one ligand has been co-crystallized with the LBD. These ligands can be naturally overlaid in space by superimposing the surrounding binding pockets. Following such superposition, the collective location of atoms of specific types in space presents valuable information that can be used in docking-based prediction and characterization of ligand activities. Specifically, compounds will be flexibly docked to and scored by the chemical property fields derived from the experimental ligand positions (Figure 5.7a–c).

The atomic property field (APF) method was developed fairly recently.[17] It represents the pharmacophore features of the superimposed high-affinity ligands (referred to as *APF seeds*) by seven continuous fields calculated in a three-dimensional grid. These seven fields represent the following properties: hydrogen bond donor and acceptor potential, sp^2 vs sp^3 hybridization, lypophilicity, size, charge and electronegativity. Each atom can contribute to multiple properties and the property peaks are smoothed in 3D space according to a Gaussian distribution with a given radius (Figure 5.7a). The radius of 1.4 Å was used in the docking simulations presented here. The multiple APF seeds may contribute differently to the APF fields depending on their experimentally measured binding affinities, pharmacokinetic properties, *etc.*

The cumulative 3D property fields can be used for docking and scoring of new compounds in exactly the same manner as receptor grid potential fields are used in docking to the receptor pocket.[36] However, the target function to be optimized now has to include the intramolecular energy of the compound in combination with the APF-fit energy calculated from the cumulative APF fields. Thorough sampling of each compound in the 3D space results in an energetically favorable conformation that best fits the atomic property fields and its APF fit score can be used as a predictor of the activity of the compound against the receptor in question.

To illustrate the utility of this approach, we constructed APF fields for all 25 NRs with available 3D seeds in the PDB. The multiple copies of identical chemicals were excluded from the process to avoid overrepresentation of selected chemical scaffolds. For simplicity, each unique crystallographic ligand was represented with an equal weight. The models obtained were validated in retrospective screening for known high-affinity modulators of the target

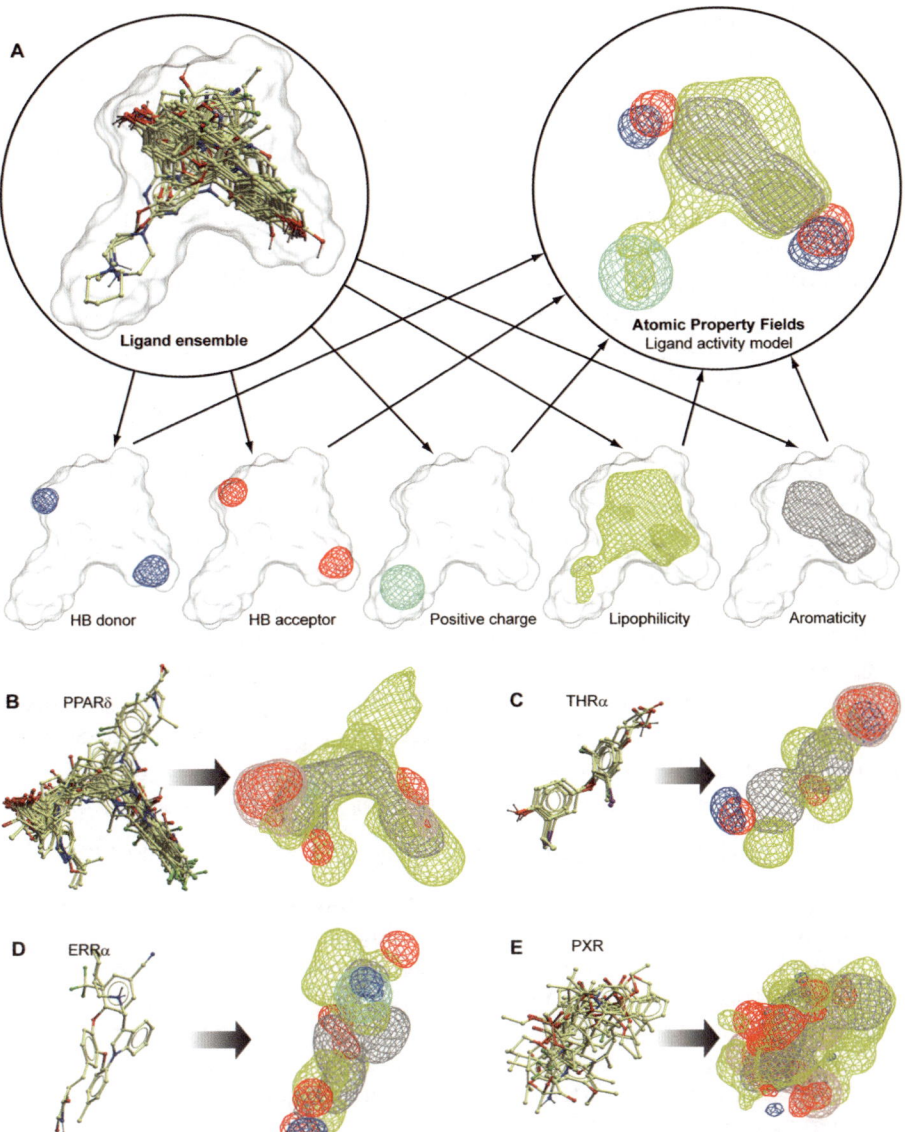

Figure 5.7   Atomic property fields. (a) The construction of APF fields (for a detailed description and numerical parameters, see ref. 17). Briefly, seven 3D grid potentials are calculated for the ligand ensemble representing the pharmacophore features of ligand atoms. Empty grids are not shown. (b, c) Representative NR ligand ensembles with well-defined pharmacophore fields. (d) The ERRα pharmacophore field is poorly defined owing to the small number and diverse chemistry and binding modes of the co-crystallized ligands. (e) The PXR pharmacophore field is poorly defined owing to high chemical diversity of its modulators.

receptors among property-matched decoys (see Section 5.4.1.1 for a more detailed description of active and decoy molecule sets). This exercise resulted in very high recognition performance towards known actives in 19 out of 25 cases (Figure 5.6). In most cases, the selectivity of the APF model even exceeded that of the traditional receptor ensemble docking, while taking only a small fraction of CPU time.

This screen, however, also highlighted the drawbacks of the APF approach and the limits of its applicability. In particular, it showed that when there are only a few seed ligands and the seed ligands are chemically distinct from the majority of active compounds, the APF model may not be capable of providing any recognition (exemplified by the case of ERRα, Figure 5.7d). Also, wide chemical variations between the ligands and a large volume of the binding pocket resulted in a poorly defined APF field, which was not selective towards known actives (*e.g.*, PXR, Figure 5.7e).

As is, the APF docking approach does not penalize the bulky compound fragments that do not fit into the APF density, provided that the core aligns well. Consequently, the superstructures of the active compounds score at least as well as the active compounds themselves, although in reality they may be too bulky even to fit sterically in the pocket. To address this issue, one needs to introduce an additional grid potential in APF docking that represents the prohibited regions in space, the so-called *excluded volume*. Adding this component is expected to improve the APF screening performance in realistic settings, where the properties of the inactive compounds do not necessarily match with those of the actives.

5.4.3 Prediction of Ligand Selectivity and Polypharmacology

The importance of finding subtype-selective modulators for NRs cannot be overestimated. The effects mediated by binding of drugs to the close (or not so close) homologs of their target receptors adds a new level of complexity to the profiles of action in the human body (*e.g.*, ref. 37). The 3D receptor-based and ligand-based models described above can be used not only to predict high-affinity modulators for a given receptor, but also to derive relative propensities of a single compound to different NRs. The additional challenge is created by the fact that here the active compounds have to be distinguished not from property-matched decoys, but from actives for another, sometimes closely related, receptor.

The binding scores calculated for each ligand and each receptor model as described above may provide some information about the relative affinity. However, although the scores can capture the relative ranking of different compounds for the same receptor well, they are not always accurate enough to represent the absolute binding free energy. An attempt was made to use the raw scores for 19 models of different NRs for selectivity profiling.[38] Although a certain level of success was achieved, the raw score against single models had only limited discrimination power. The reasons for non-ideal discrimination can be narrowed down to the following contributions: (i) the protein-dependent part of the binding energy that is not included in the docking score, *e.g.*, the protein-dependent entropy loss upon ligand binding; (ii) a pocket conformation – the accuracy of different models may be different and systematic geometric errors or features may shift the binding score consistently up or down; (iii) other errors or simplifications of the pocket model including partial atomic charges, individual strength of hydrogen bonds, *etc.*

The systematic protein-dependent score shift can be taken into account by introduction of score offsets. The distribution of the scores from a large decoy set is calculated and shifted so that the edge of the noise distribution corresponds to the zero binding score. The offset was defined as the 5th percentile of the best scores from the decoy distribution, where the decoys were defined as the whole database of nuclear receptor ligands.[16] These naïve offsets improved ligand–receptor recognition by the shifted score. The offset can be further optimized by a more detailed fitting algorithm, but for the NRs considered,[16] it made relatively small changes in the offset. The shifted scores reduced the false positives coming generally from the better scoring pockets. The same method can further be applied to fine-tune individual scores coming from different conformations of the same protein.

The ligand-based 3D activity models described above can also be used for compound profiling. To illustrate their ability to predict selective compounds, we chose two sets of closely related NR subtypes: peroxisome proliferator-activated receptors (PPARs) α, δ and γ and NR3C family receptors, androgen, progesterone, glucocorticoid and mineralocorticoid receptors. Within both classes, substantial cross-reactivity is observed, as shown in Figure 5.2b. In each case, the selectivity of compounds targeting one subtype was distinguished from subtype-selective modulators of others. Figure 5.8 shows that a substantial initial enrichment is observed when the PPARδ model is used to screen for PPARδ-selective compounds, but not when PPARα or PPARγ models are used instead. On the contrary, PPARγ-selective ligands are clearly more accurately recognized by the PPARγ model than by PPARα or δ models. As with any other method, APF has its weaknesses. For example, all related subtype models appear to have approximately the same level of selectivity when an attempt is made to identify PR-selective ligands or PPARα-selective ligands.

The score offset approach is not applicable to the APF docking because APF scores are calculated on a different scale. To yield an absolute APF-based compound activity and selectivity profile predictor, the scores need to be further normalized. One way of doing it is to analyze the score distribution for decoy/inactive compounds and assign each score with a probability of observing this score for an inactive compound. As described above (Section 5.4.1.1), fitting the decoy score distribution into an analytical function (Gaussian or extreme value distribution function) with subsequent evaluation of parameters of this function provides a compact and accurate way of accounting for protein-specific binding score offsets. For each new compound,

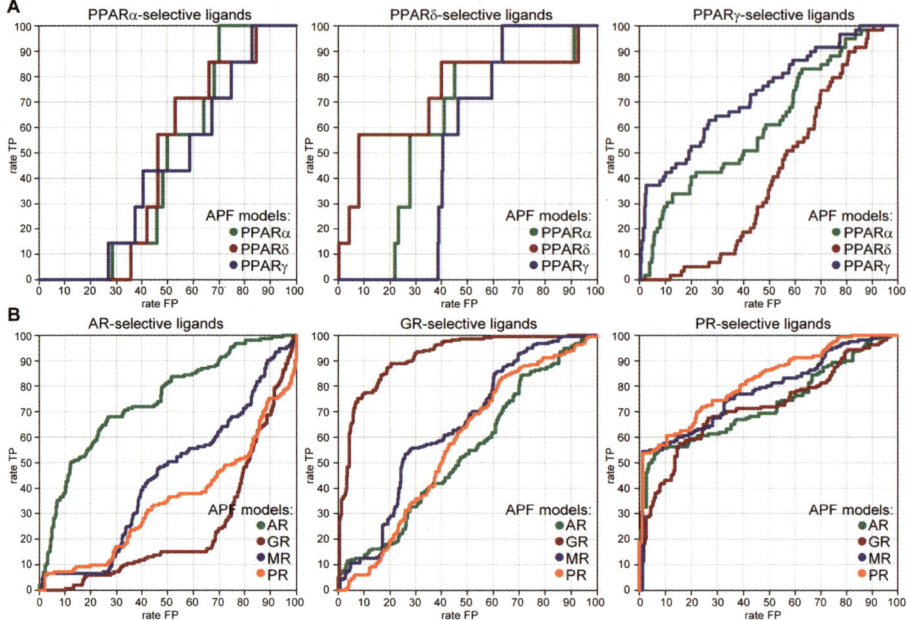

Figure 5.8 Prediction of ligand selectivity by the APF models of (a) NR1C and (b) NR3C group receptors. The combined chemical sets consisting of all subtype-selective modulators for each group were docked into the APF models for all receptors. For two of three NR1C group receptors (PPARδ and PPARγ), the subtype-selective ligands, but not other ligands, are recognized by the cognate APF model. A similar situation is observed for two of four NR3C group receptors, AR and GR. Prediction of PPARα- and PR-selective compounds remains a challenging task for the APF models. Owing to an insufficient number of MR-selective compounds, MR data are not shown.

a linear transformation of a raw docking score followed by a *p*-value evaluation provides an activity call for a ligand–receptor pair.

Acknowledgements

We thank Maxim Totrov for discussions about the ICM tools for docking, scoring and atomic property fields. We thank Fiona McRobb for help with chemistry set compilation, reading the manuscript and useful comments about estrogen receptor-related issues. This work was supported in part by NIH grants R01 GM071872, U01 GM094612 and U54 GM094618.

References

1. J. P. Overington, B. Al-Lazikani and A. L. Hopkins, How many drug targets are there?, *Nat. Rev. Drug Discov.*, 2006, **5**, 993–996.

2. F. Gruen and B. Blumberg, Environmental obesogens: organotins and endocrine disruption via nuclear receptor signaling, *Endocrinology*, 2006, **147**, s50–s55.

3. C. Casals-Casas and B. Desvergne, Endocrine disruptors: from endocrine to metabolic disruption, *Annu. Rev. Physiol.*, 2011, **73**, 135–162.

4. A. R. Clark and M. G. Belvisi, Maps and legends: the quest for dissociated ligands of the glucocorticoid receptor, *Pharmacol. Ther.*, 2012, **134**, 54–67.

5. D. P. McDonnell and S. E. Wardell, The molecular mechanisms underlying the pharmacological actions of ER modulators: implications for new drug discovery in breast cancer, *Curr. Opin. Pharmacol.*, 2010, **10**, 620–628.

6. J. O. Jones, Improving selective androgen receptor modulator discovery and preclinical evaluation, *Expert Opin. Drug Discov.*, 2009, **4**, 981–993.

7. K. De Bosscher, Selective glucocorticoid receptor modulators, *J. Steroid Biochem. Mol. Biol.*, 2009, **120**, 96–104.

8. K. Chwalisz, M. C. Perez, D. DeManno, C. Winkel, G. Schubert and W. Elger, Selective progesterone receptor modulator development and use in the treatment of leiomyomata and endometriosis, *Endocrine Rev.*, 2005, **26**, 423–438.

9. W. Wharton, E. Hirshman, P. Merritt, L. Doyle, S. Paris and C. Gleason, Oral contraceptives and androgenicity: influences on visuospatial task performance in younger individuals, *Exp. Clin. Psychopharmacol.*, 2008, **16**, 156–164.

10. R. Sitruk-Ware and A. Nath, Metabolic effects of contraceptive steroids. *Rev. Endocrine Metab. Disord.*, 2011, **12**, 63–75.

11. M. Schapira, B. M. Raaka, H. H. Samuels and R. Abagyan, *In silico* discovery of novel retinoic acid receptor agonist structures, *BMC Struct. Biol.*, 2001, **1**, 1–1.

12. M. Schapira, B. M. Raaka, H. H. Samuels and R. Abagyan, Rational discovery of novel nuclear hormone receptor antagonists, *Proc. Natl. Acad. Sci. U. S. A.*, 2000, **97**, 1008–1013.

13. M. Schapira, B. M. Raaka, S. Das, L. Fan, M. Totrov, Z. Zhou, S. R. Wilson, R. Abagyan and H. H. Samuels, Discovery of diverse thyroid hormone receptor antagonists by high-throughput docking, *Proc. Natl. Acad. Sci. U. S. A.*, 2003, **100**, 7354–7359.

14. S. Kortagere, M. D. Krasowski, E. J. Reschly, M. Venkatesh, S. Mani and S. Ekins, Evaluation of computational docking to identify pregnane X receptor agonists in the ToxCast database, *Environ. Health Perspect.*, 2010, **118**, 1412–1417.

15. W. H. Bisson, A. V. Cheltsov, N. Bruey-Sedano, B. Lin, J. Chen, N. Goldberger, L. T. May, A. Christopoulos, J. T. Dalton, P. M. Sexton, X. K. Zhang and R. Abagyan, Discovery of antiandrogen activity of nonsteroidal scaffolds of marketed drugs, *Proc. Natl. Acad. Sci. U. S. A.*, 2007, **104**, 11927–11932.

16. S.-J. Park, I. Kufareva and R. Abagyan, Improved docking, screening and selectivity prediction for small molecule nuclear receptor modulators using conformational ensembles, *J. Comput.-Aided Mol. Des.*, 2010, **24**, 459–471.

17. M. Totrov, Atomic property fields: generalized 3D pharmacophoric potential for automated ligand superposition, pharmacophore elucidation and 3D QSAR, *Chem. Biol. Drug Des.*, 2008, **71**, 15–27.

18. P. Germain, B. Staels, C. Dacquet, M. Spedding and V. Laudet, Overview of nomenclature of nuclear receptors, *Pharmacol, Rev.*, 2006, **58**, 685–704.

19. Y. Zhan, X. Du, H. Chen, J. Liu, B. Zhao, D. Huang, G. Li, Q. Xu, M. Zhang, B. C. Weimer, D. Chen, Z. Cheng, L. Zhang, Q. Li, S. Li, Z. Zheng, S. Song, Y. Huang, Z. Ye, W. Su, S.-C. Lin, Y. Shen and Q. Wu, Cytosporone B is an agonist for nuclear orphan receptor Nur77, *Nat. Chem. Biol.*, 2008, **4**, 548–556.

20. J.-j. Liu, H.-n. Zeng, L.-r. Zhang, Y.-y. Zhan, Y. Chen, Y. Wang, J. Wang, S.-h. Xiang, W.-j. Liu, W.-j. Wang, H.-z. Chen, Y.-m. Shen, W.-j. Su, P.-q. Huang, H.-k. Zhang and Q. Wu, A unique pharmacophore for activation of the nuclear orphan receptor Nur77 *in vivo* and *in vitro*, *Cancer Res.*, 2010, **70**, 3628–3637.

21. K. T. Nguyen, L. C. Blum, R. van Deursen and J.-L. Reymond, Classification of organic molecules by molecular quantum numbers, *ChemMedChem*, 2009, **4**, 1803–1805.

22. M. A. Gallo and D. Kaufman, Antagonistic and agonistic effects of tamoxifen: significance in human cancer, *Semin. Oncol.*, 1997, **24**, 1–71.

23. A. K. Shiau, D. Barstad, J. T. Radek, M. J. Meyers, K. W. Nettles, B. S. Katzenellenbogen, J. A. Katzenellenbogen, D. A. Agard and G. L. Greene, Structural characterization of a subtype-selective ligand reveals a novel mode of estrogen receptor antagonism, *Nat. Struct. Mol. Biol.*, 2002, **9**, 359–364.

24. M. J. Keiser, V. Setola, J. J. Irwin, C. Laggner, A. I. Abbas, S. J. Hufeisen, N. H. Jensen, M. B. Kuijer, R. C. Matos, T. B. Tran, R. Whaley, R. A. Glennon, J. Hert, K. L. H. Thomas, D. D. Edwards, B. K. Shoichet and B. L. Roth, Predicting new molecular targets for known drugs, *Nature*, 2009, **462**, 175–181.

25. I. Kufareva, A. V. Ilatovskiy and R. Abagyan, Pocketome: an encyclopedia of small-molecule binding sites in 4D, *Nucleic Acids Res.*, 2012, **40**, 535–540.

26. H. C. A. Raaijmakers, J. E. Versteegh and J. C. M. Uitdehaag, The X-ray structure of RU486 bound to the progesterone receptor in a destabilized agonistic conformation, *J. Biol. Chem.*, 2009, **284**, 19572–19579.

27. M. Totrov and R. Abagyan, Flexible protein-ligand docking by global energy optimization in internal coordinates, *Proteins*, 1997, Suppl 1, 215–220.

28. M. Totrov and R. Abagyan, Flexible ligand docking to multiple receptor conformations: a practical alternative, *Curr. Opin. Struct. Biol.*, 2008, **18**, 178–184.

29. G. Bottegoni, I. Kufareva, M. Totrov and R. Abagyan, Four-dimensional docking: a fast and accurate account of discrete receptor flexibility in ligand docking, *J. Med. Chem.*, 2009, **52**, 397–406.
30. N. Huang, B. K. Shoichet and J. J. Irwin, Benchmarking sets for molecular docking, *J. Med. Chem.*, 2006, **49**, 6789–6801.
31. M. Olah, R. Rad, L. Ostopovici, A. Bora, N. Hadaruga, D. Hadaruga, R. Moldovan, A. Fulias, M. Mractc and T. I. Oprea, WOMBAT and WOMBAT-PK: bioactivity databases for lead and drug discovery, in *Chemical Biology: from Small Molecules to Systems Biology and Drug Design*, ed. S. L. Schreiber, T. M. Kapoor and G. Wess, Wiley-VCH, Weinheim, 2008, pp. 760–786.
32. M. A. C. Neves, M. Totrov and R. Abagyan, Docking and scoring with ICM: the benchmarking results and strategies for improvement, *J. Comput.-Aided Mol. Des.*, 2012, May 9. [Epub ahead of print]
33. A. Gaulton, L. J. Bellis, A. P. Bento, J. Chambers, M. Davies, A. Hersey, Y. Light, S. McGlinchey, D. Michalovich, B. Al-Lazikani and J. P. Overington, ChEMBL: a large-scale bioactivity database for drug discovery, *Nucleic Acids Res.*, 2011, **40**(DI), D1100–D1107.
34. A. J. W. Orry and R. Abagyan, Preparation and refinement of model protein–ligand complexes, *Methods Mol. Biol.*, 2012, **857**, 351–373.
35. W. H. Bisson, A. V. Cheltsov, N. Bruey-Sedano, B. Lin, J. Chen, N. Goldberger, L. T. May, A. Christopoulos, J. T. Dalton, P. M. Sexton, X. K. Zhang and R. Abagyan, Discovery of antiandrogen activity of nonsteroidal scaffolds of marketed drugs, *Proc. Natl. Acad. Sci. U. S. A.*, 2007, **104**, 11927–11932.
36. M. Totrov and R. Abagyan, Derivation of sensitive discrimination potential for virtual ligand screening, in *Proceedings of the Third Annual International Conference on Computational Molecular Biology*, ACM, Lyon, 1999,.
37. D. C. Leitman, S. Paruthiyil, O. I. Vivar, E. F. Saunier, C. B. Herber, I. Cohen, M. Tagliaferri and T. P. Speed, Regulation of specific target genes and biological responses by estrogen receptor subtype agonists, *Curr. Opin. Pharmacol.*, 2010, **10**, 629–636.
38. M. Schapira, R. Abagyan and M. Totrov, Nuclear hormone receptor targeted virtual screening, *J. Med. Chem.*, 2003, **46**, 3045–3059.

CHAPTER 6

Quantum Chemical Studies of Estrogenic Compounds

WAYNE B. BOSMA*[a] AND MICHAEL APPELL[b]

[a] Department of Chemistry and Biochemistry, Bradley University, Peoria, IL 61625, USA; [b] Bacterial Foodborne Pathogens and Mycology Research, United States Department of Agriculture (USDA), Agricultural Research Service, National Center for Agricultural Utilization Research, 1815 N. University Street, Peoria, IL 61604, USA
*E-mail: bosma@bradley.edu

6.1 Introduction

Nuclear receptors play an important role in the regulation of gene expression and gene function in animals through the direct interaction with DNA, especially during development and homeostasis maintenance.[1] Activation of this class of receptors occurs through the molecular recognition with a broad range of hydrophobic ligands depending on the type of nuclear receptor. In addition, orphan receptors have been observed that share many characteristics of nuclear receptors without known endogenous ligands.[2]

The ancestral function of evolutionary precursors to nuclear receptors has been proposed to be that of biological sensors for hydrophobic molecules, such as retinoids, steroids and fatty acids.[3] These receptors are suggested to have evolved their high nanomolar-range affinity for steroids and other bioactives, enabling them to be useful for a range of activities, including regulation of gene expression. The rational design of nuclear receptor bioactives is complicated by nuclear receptors exhibiting affinity for a variety of endogenous ligands, affinity for xenoestrogens and the presence of orphan receptors. In addition,

RSC Drug Discovery Series No. 30
Computational Approaches to Nuclear Receptors
Edited by Pietro Cozzini and Glen E. Kellogg
© The Royal Society of Chemistry 2012
Published by the Royal Society of Chemistry, www.rsc.org

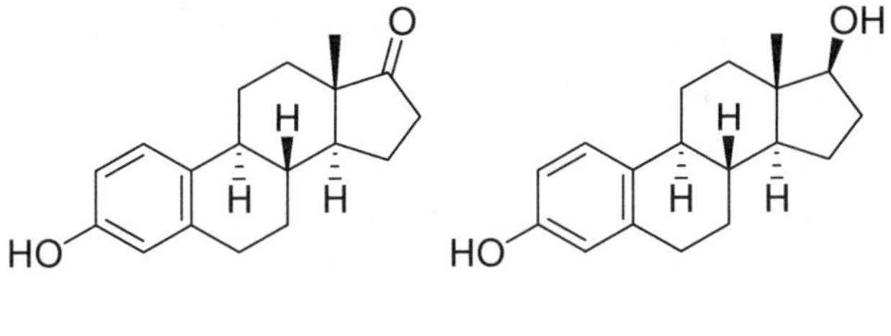

Figure 6.1   Structures of estrone (**1**) and β-estradiol (**2**).

several factors influence molecular recognition, including the flexibility of the target receptors.[4]

Recent structure–function studies have helped to elucidate some general trends associated with ligands exhibiting high affinity for nuclear receptors, including the presence of aromatic ring systems and a degree of hydrophobicity.[5] The structures of substrates for the nuclear receptors are known and quantum chemical methods may be used to provide insight into the observed structure–activity relationships for these known ligands. In this chapter, quantum chemical studies and other *in silico* investigations of the natural estrognic compounds estrone (**1**) and β-estradiol (**2**) (Figure 6.1) are reviewed. In addition, we touch on selected studies on non-steriodal estrogenic compounds (Figure 6.2). Quantum mechanical methods have provided valuable information on the molecular properties of estrogenic compounds, and also the reaction mechanisms with important biomolecules, quantitative structure–activity relationship (QSAR) studies and the binding to macromolecules and formation of inclusion complexes.

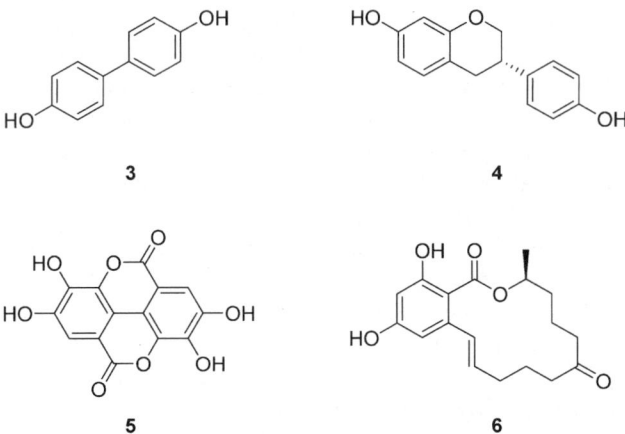

Figure 6.2   Structures of 4,4′-biphenol (**3**), equol (**4**), ellagic acid (**5**) and zearalenone (**6**).

6.2 Single-molecule Density Functional Modeling and *Ab Initio* Modeling Studies

Single-molecule studies of estrogenic compounds include studies on the endogenous estrogens estrone and estradiol and their derivatives. These studies have primarily been carried out using density functional methods that employ the B3LYP exchange-correlation functional (Becke, three-parameter, Lee–Yang–Parr).

The influence of chemical structure on the increased carcinogenicity of estradiol compared with estrone was investigated using the B3LYP functional at the 6–31G level.[6] The study explored similarities and differences in the electrostatic potentials of the molecules, and also the differences densities of electronic states. Based upon criteria shown to work in other systems, it was suggested that the reduced carcinogenicity of estrone is due to the ketone functional group of estrone reducing the aromatic character and the reactivity of the phenolic region of estrone. This region is important for the formation of the reactive anion species of the phenolic hydroxyl that is associated with carcinogenicity.

The electron density and electrostatic properties of estrone were studied using B3LYP/6–311++G(d,p) single-point calculations using experimentally obtained (crystal structure) geometries.[7] The hydroxyl and carbonyl moieties exhibit significantly negative regions in the molecular electrostatic potential. Moieties that show promise for interaction with nuclear receptors include the electronegative regions associated with the aromatic ring and the phenolic hydroxyl and hydrogen bonding association with the phenolic hydroxyl and the ketone. The calculated electron distribution and geometry were found to agree well with the corresponding experiments.

In addition, B3LYP/6–31+G* calculations assisted in the determination of the stereochemistry of a series of novel synthetic oxazaphosphorinenane estrone methyl ether derivatives possessing a heterocyclic fused ring moiety at the C-16–C-17 positions.[8] The computations were used to determine the energetic plausibility of proposed conformations and aided in the interpretation of the NMR experiments to determine the experimentally observed structures. The heterocyclic ring system exhibited an unusual distorted boat conformation as its most stable conformation.

Quantum chemical methods have provided insight into the binding properties and antioxidant activities of several non-steriodal estrogenic compounds. Hartree–Fock and density functional calculations provided insight into the antioxidant properties of 4,4′-biphenol (**3**), 2,2′-biphenol and phenol.[9] Equilibrium geometry optimization calculations were carried out using the Hartree–Fock method and the 6–31G* basis set for ground-state studies and studies of phenoxy free radicals were conducted at the UHF/6–31G* level of theory. The bond dissociation energies were calculated using electronic energies provided by B3LYP single-point energy calculations at the 6–31G* level. The ionization potential was calculated using HF/6–31G* methods. Experimental studies demonstrated that COX-2 expression in RAW cells is enhanced by 4,4'-biphenol compared with 2,2'-biphenol and phenol. It was rationalized that the increased antioxidant activities of 4,4'-biphenol over 2,2'-biphenol and phenol are associated with differences in ionization potential.

The electronic structures of polychlorinated biphenol derivatives exhibiting estrogenic activities at the estrogen receptor ERα were investigated using the B3LYP/6–311G(d,p) level of theory.[10] A relationship was found between the ERα binding by polychlorinated biphenols and the sum of charges associated with the carbons and oxygen of the phenolic ring of the polychlorinated biphenol. By contrast, the expected relationship between structural rigidity and binding affinity was shown to be a less important factor.

The natural estrogen and antioxidant equol (4) was investigated using AM1 semi-empirical geometry optimization calculations and B3LYP/3–21G calculations.[11] Properties investigated included the frontier molecular orbitals, gap energies and associated molecular orbital-based properties. Although the recognition properties could not be related to the structure, the authors concluded that both phenolic hydroxyls are important to the antioxidant properties of equol.

Ellagic acid (5) exhibits estrogenic activity at the α and β estrogenic receptors, and also antioxidant activities.[12] Ellagic acid and its derivatives were investigated using B3LYP/6–31+G** calculations to gain insight into the structure and electronic properties, including bond dissociation energies, ionization potential, proton dissociation energy, proton affinity and electron transfer energy.[13] Comparison of the bond dissociation energies with those of other popular antioxidants predicts ellagic acid to be one of the most potent natural antioxidants and its mechanism of action is through the hydrogen atom transfer mechanism.

6.3 Biomolecular Studies

The chemical reactivities of hydroxylated estrone and estradiol with guanine were studied at the B3LYP/6–31G(d,p) level of theory.[14] These catechol estrogens are formed through metabolic pathways and form adducts with the N-7 of guanine. Comparing estrone and estradiol, there was no difference in the calculated free energies of reaction of the two carcinogens, meaning that the rate constants for adduct formation are essentially identical. The observed difference in reactivity is likely associated with specific molecular recognition properties of DNA and the presence of the hydroxyl in β-estradiol. Other factors to explain the difference in estrone and estradiol activities include preference in depurination reactions, selective metabolism to the reactive oxidative state through the P450 enzyme or selective detoxification of the oxidative state. Biochemical differences between the actual nuclear receptors and the guanine model for DNA may also play a role.

Mechanistic studies of how catechol estrogenic 3,4-quinones form Michael addition products with deoxyguanosine were undertaken both experimentally and computationally.[15] The computational results supported the mechanism that involves an α-keto–enol ring system intermediate in a proton-assisted Michael addition, followed by slow loss of the proton at the C-1 to restore the aromatic ring system and then cleavage of the glycosidic bond to form the estrogen–deoxyguinosine adduct.

The mechanism of action of sulfotransferase with 3′-phosphoadenosine 5′-phosphosulfate as the sulfate donor and estradiol as the sulfate acceptor was investigated using density functional, *ab initio* and semi-empirical methods.[16] The mechanism of the reaction is proposed to be complex and proceeds through formation of the neutral species of 3′-phosphoadenosine 5′-phosphosulfate, followed by S–O bond stretching, and finally through a sulfydryl transfer involving a histidine residue.

6.4   Chemical Properties

There has been a great increase in the use of quantum descriptors to develop QSAR models.[17] The use of estrogenic food additives is a serious concern and computational methods have been applied to predict the estrogenic potential of 31 candidates from a database of 1500 common food additives.[18] The computational screening was conducted using the docking program GOLD and the HINT scoring function. These methods are successful at predicting the activities of protein interactions with similar small organics, proteins and DNA. The food additives propyl gallate and 4-hexylresorcinol were identified as potential xenoestrogens.

A QSAR study was conducted on the estrogenic activities of a series of terpenoid esters from *Ferula* plants.[19] Physicochemical descriptors were determined at the B3LYP/6–31G(d,p) level and a genetic algorithm was used to relate the properties. A broad range of parameters were found to be important, including the surface polarity and the energy of the highest occupied molecular orbital (HOMO).

Recently, the relationship between electronic properties and anticoagulation biological activities of estrogens modified at the C-17 position by amino acids was investigated using the B3LYP density functional.[20] Calculations were carried out at the B3LYP/6–311G(d,p) level using B3LYP/6–31G(d) geometries to determine physicochemical descriptors, including hardness, electrophilicity and aromatic properties. It is reported that derivation at the C-17 position rendered the estrogen derivative more exposed to nucleophilic and electrophilic reaction and did not influence the acidity of the phenolic hydroxyl.

A 25-compound QSAR study which included quantum chemical descriptors was developed to describe binding affinity for the estrogenic receptor. Compounds studied included industrial estrogens and three natural estrogens, estradiol, estrone and estratriol. The best model developed was based on a steric component (molecular volume V_m) and the energy of the HOMOs (ε_{HOMO}).[21]

6.5   Complexations

Docking studies of several estrogenic and anti-estrogenic ligands with ERα and ERβ were used to examine whether molecules similar to these show any promise as alternatives to conventional hormone replacement therapy drugs.[22] Several different docking methods were used and the HINT (hydropathic interaction) software was employed to score the results. The study made progress towards the development of scoring functions that can effectively predict binding efficiency in systems where crystal structure data are not available.

β-(1,3)-D-Glucans from *Saccharomyces cerevisiae* are able to sequester the estrogenic toxin zearalenone (**6**) and other mycotoxins and empirical modeling has provided insight into the mechanisms.[23,24] Calculations using the CFF91 force field indicate that zearalenone binds to single helical β-(1,3)-D-glucans through exceptional steric fit and hydrogen bond interactions between the β-(1,3)-D-glucan hydroxyls and the phenolic hydroxyls and ketone group of zearalenone.

Cyclodextrins are cyclic glucopyranoses capable of forming guest–host complexes with hydrophobic small organic molecules, including the estrogenic mycotoxin zearalenone. β-Cyclodextrin–zearalenone complexes have been investigated using PM3 semi-empirical quantum mechanical methods and suggest that zearalenone can form several types of complexes with β-cyclodextrins with the ketone and resorcylic acid moieties encapsulated by the cyclodextrin.[25] Complexation interactions of cyclodextrins with the estrogenic mycotoxin zearalenone and other mycotoxins have been investigated using the HINT force field.[26,27] The β-cyclodextrin:zearalenone ratio was observed to be 1:1 by NMR studies. However, the intense fluorescence enhancement of β-cyclodextrin–zearalenone complexes and molecular modeling may suggest a 2:1 complex, with the resorcylic acid moiety encapsulated. β-Cyclodextrin complexes with structurally related α- and β-zearalenol were formed with hydrogen bonding between the oxidryl moiety and the hydroxyls of the cyclodextrin.

Acknowledgements

Mention of trade names or commercial products in this chapter is solely for the purpose of providing specific information and does not imply recommendation or endorsement by the US Department of Agriculture. USDA is an equal opportunity provider and employer.

References

1. J. R. Olefsky, *J. Biol. Chem.*, 2001, **276**(40), 36863.
2. J. T. Moore, J. L. Collins and K. H. Pearce, *ChemMedChem*, 2006, **1**, 504.
3. G. V. Markov and V. Laudet, *Mol. Cell. Endocrinol.*, 2011, **334**, 21.
4. P. Cozzini, G. E. Kellogg, F. Spyrakis, D. J. Abraham, G. Costantino, A. Emerson, F. Fanelli, H. Gohlke, L. A. Kuhn, G. M. Morris, M. Orozco, T. A. Pertinhez, M. Rizzi and C. A. Sotriffer, *J. Med. Chem.*, 2008, **51**, 6237.

5. D. L. Bain, A. F. Heneghan, K. D. Connaghan-Jones and M. T. Miura, *Annu. Rev. Physiol.*, 2007, **69**, 201.

6. A. Picazo and R. Salcedo, *J. Mol. Struct.: THEOCHEM*, 2003, **624**, 29.

7. EA. Zhurova, C. F. Matta, N. Wu, V. V. Zhurov and A. A. Pinkerton, *J. Am. Chem. Soc.*, 2006, **128**, 8849.

8. É. Frank, B. Kazi, Z. Mucsi, K. Ludányi and G. Keglevich, *Steriods*, 2007, **72**, 446.

9. Y. Murakami, H. Ishii, S. Hoshina, N. Takada, A. Ueki, S. Tanaka, Y. Kadoma, S. Ito, M. Machino and S. Fujisawa, *Anticancer Res.*, 2009, **29**, 2403.

10. S. Arulmozhiraja, F. Shiraishi, T. Okumura, M. Iida, H. Takigami, J. S. Edmonds and M. Morita, *Toxicol. Sci.*, 2005, **84**, 49

11. F. Erkoc, M. Yilmazer and S. Erkoc, *J. Mol. Struct.:THEOCHEM*, 2005, **713**, 37.

12. M. Larrosa, A. González-Sarrías, M. T. García-Conesa, F. A. Tomás-Barberán and J. C. Espín, *J. Agric. Food Chem.*, 2000, **54**, 1611.

13. J. Zhang, Y. Xiong, B. Peng, H. Gao and Z. Zhou, *Comput. Theor. Chem.*, 2011, **963**, 148.

14. P. Huetz, E. E. Kamarulzaman, H. A. Wahab and J. Mavri, *J. Chem. Inf. Comput. Sci.*, 2004, **44**, 310.

15. D. E. Stack, G. Li, A. Hill and N. Hoffman, *Chem. Res. Toxicol.*, 2008, **21**, 1415.

16. L. Bartolotti, Y. Kakuta, L. Pedersen, M. Negishi and L. Pedersen, *J. Mol. Struct.: THEOCHEM*, 1999, **461–462**, 105.

17. P. G. De Benedetti and F. Fanelli, *Drug Discov. Today*, 2010, **15**(19/20), 859.

18. A. Amadasi, A. Mozzarelli, C. Meda, A. Maggi and P. Cozzini, *Chem. Res. Toxicol.*, 2009, **22**, 52.

19. B. F. Rasulev, A. I. Saidkhodzhaev, S. S. Nazrullaev, K. S. Akhmedkhodzhaeva, Z. A. Khushbaktova and J. Leszczynski, *SAR QSAR Environ. Res.*, 2007, **18**, 663.

20. A. Raya, C. Barrientos-Salcedo, C. Rubio-Poo and C. Soriano-Correa, *Eur. J. Med. Chem.*, 2011, **46**, 2463.

21. J. Y. Hu and T. Aizawa, *Water Res.*, 2003, **37**(6), 1213.

22. P. Cozzini and T. Dottorini, *Eur. J. Med. Chem.*, 2004, **39**, 601.

23. A. Yiannikouris, G. André, A. Buléon, G. Jeminet, I. Canet, J. F. Gérard Bertin and J.-P. Jouany, *Biomacromolecules*, 2004, **5**, 2176.

24. A. Yiannikouris, G. André, L. Poughon, J. François, C.-G. Dussap, G. Jeminet, G. Bertin and J.-P. Jouany, *Biomacromolecules*, 2004, **7**, 1147.

25. M. Appell and C. M. Maragos, in *Mycotoxin Prevention and Control in Agriculture*, ed. M. Appell, D. F. Kendra and M. W. Trucksess, ACS Symposium Series, Vol. 1031, American Chemical Society, Washington, DC, 2009, p. 293.

26. P. Cozzini, G. Ingletto, R. Singh and C. Dall'Asta. *Int. J. Mol. Sci.*, 2008, **9**(12), 2474.

27. A. Amadasi, C. Dall'Asta, G. Ingletto, R. Pela, R. Marchelli and P. Cozzini, *Bioorg. Med. Chem.*, 2007, **15**, 4585.

CHAPTER 7

A Nuclear G Protein-coupled Estrogen Receptor, GPER. Homology Modeling Studies Toward Its Ligand-binding Mode Characterization

CHRISTOPHER K. ARNATT AND YAN ZHANG*

Department of Medicinal Chemistry, Virginia Commonwealth University, Richmond, VA 2329, USA
*E-mail: yzhang2@vcu.edu

7.1 G Protein Estrogen Receptor

7.1.1 Identification

Estrogens serve a variety of functions within the body, ranging from maintaining the female reproduction system to aiding in neuroprotection.[1] They have also been shown to play multiple roles in diseases such as cancer. More specifically, 17β-estradiol (Figure 7.1) has been shown to be a mediator in breast cancer and a number of other estrogen-sensitive cancers.[1] Currently, there are several cancer therapeutics which target 17β-estradiol signaling, for example, tamoxifen, fulvestrant and raloxifene.[1]

Historically, there have been two classical cytoplasmic/nuclear estrogen receptors connected to 17β-estradiol's effects in the body. These two receptors,

RSC Drug Discovery Series No. 30
Computational Approaches to Nuclear Receptors
Edited by Pietro Cozzini and Glen E. Kellogg
© The Royal Society of Chemistry 2012
Published by the Royal Society of Chemistry, www.rsc.org

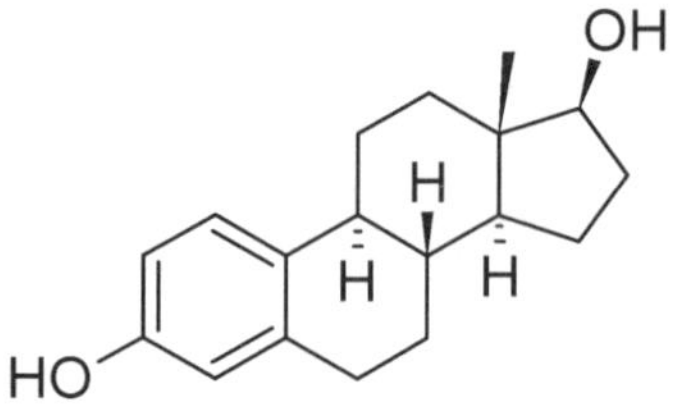

Figure 7.1 Structure of the G protein estrogen receptor (GPER) agonist 17β-estradiol.

estrogen receptor α and β (ERα and ERβ), primarily induce signaling at the genomic level, but have also been linked to rapid non-genomic signaling under certain circumstances.[2] Further evidence for 17β-estradiol's involvement in rapid non-genomic signaling arose when a study showed that a gene encoding for a GPCR was over-expressed in certain breast cancer cell lines.[3] This GPCR, now named G protein estrogen receptor (GPER, GPR30), was considered an orphan receptor until additional studies proved that the downstream signaling was induced by estrogen binding. It was observed that 17β-estradiol could mediate the activation of ERK1/2 through GPER in SKBR3 breast cancer cells which endogenously do not express ERα and ERβ.[4] Any response to 17β-estradiol in those cells would therefore have to occur through a mechanism other than acting through classical estrogen receptors. It is important to note that GPER can be, and often is, expressed along with ERα and ERβ in normal tissues contributing to the complex signal transduction caused by estrogens.[1,5,6] Several intracellular signaling pathways are induced by GPER upon stimulation by estrogens: activation of ERK via the transactivation of EGFR, cAMP elevation, calcium mobilization and regulation of gene expression.[7]

7.1.2 Function

Although the specific signaling of GPER has been studied in several systems, its exact physiological functions are still being elucidated.[6] As an estrogen receptor, its role in the reproductive system is expected; however, it has also been found to be involved with the nervous system, immune system, cardiovascular system, renal system, pancreatic function, glucose metabolism and bone growth.[8] Within all of those potential functions, the most promising therapeutic direction is treating estrogen-sensitive cancers. Clinically, GPER has been shown to be expressed in over 50% of breast carcinomas and its expression has been correlated with increased tumor size, metastasis and poor clinical outcomes.[9–12] GPER is also expressed in tumors of the prostate, lung, endometrium, ovaries, thyroid and testes.[8,10]

Several breast cancer treatments, particularly tamoxifen and fulvestrant, while being antagonists of ERα and ERβ act also as GPER agonists. Particularly, the therapeutic function of tamoxifen is normally to act as an estrogen receptor antagonist and inhibit the effects that estrogens have in promoting cancer cell growth in estrogen receptor-positive carcinomas.[2,13] Recently, it has been found also to act as a GPER agonist and stimulate GTPγS binding and adenylate cyclase activity.[2] Therefore, the agonism of tamoxifen on GPER in those same tissues may represent a novel mechanism for tamoxifen resistance in certain breast cancers.[13]

7.1.3 Recent Progress

Although the complex role of GPER in breast cancer has not been fully characterized, its ability for predicting clinical outcomes has shown the importance of developing therapeutic agents that target it. In order to elucidate the function and therapeutic potential of GPER, the Prossnitz group at the University of New Mexico discovered both a selective agonist and an antagonist by utilizing virtual screening for 17β-estradiol-like compounds in a large molecule library (Figure 7.2).[14,15] The selective agonist G-1 has a K_i of 11 nM for GPER and does not bind to ERα or ERβ at concentrations up to 1 μM.[14] A selective antagonist, G-15, was found in a similar manner through virtual screening while G-1 was used as the reference compound instead of 17β-estradiol. G-15 binds GPER with a K_i of 20 nM and does not bind ERα or ERβ at concentrations up to 10 μM, although later studies have shown some activity through ERα and ERβ by this compound.[15,16] In calcium mobilization assays, G-15 was also found to inhibit calcium ion mobilization mediated by both 17β-estradiol and G-1.[15] Further studies revealed an additional selective antagonist, G-36, with decreased activity at ERα and ERβ. This new ligand is based on the same scaffold as the two previous ligands but has an additional, bulky and lipophilic group that is believed to be responsible for the lower binding affinity at ERα and ERβ.[16]

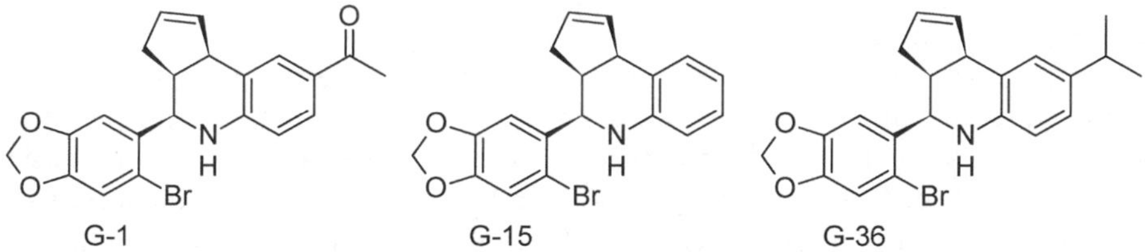

Figure 7.2 Structure of GPER-selective agonist (G-1) and antagonists (G-15, G-36).

7.1.4 Basis as a Nuclear Receptor

An increasing number of studies have begun to reveal that some GPCRs can localize and signal from the nuclear membrane in addition to the cellular membrane.[17] Specifically, angiotensin II type I and type II receptors,[18–21] prostaglandin receptors,[22] metabotropic glutamate receptors,[23] endothelin receptors,[24] lysophosphatidic acid receptors,[25] apelin receptors,[20] α_1-adrenergic receptors,[26] bradykinin B_2 receptors,[20] platelet-activating factor receptors,[27] β-adrenergic receptors,[28] and GPER[29,30] have all been shown to mediate signaling from the nuclear membrane.

Two main studies using biochemical and chemical techniques have shown that GPER can mediate rapid intracellular signaling. The first method utilized a green fluorescent protein (GFP)-tagged GPER conjugate along with a fluorescent form of 17β-estradiol through specifically monitoring calcium mobilization caused by the tagged GPER.[29] The second study employed cell-permeable and -impermeable estrogen derivatives to access which could stimulate GPER-specific signaling.[30] In both cases, it was shown that GPER was expressed intracellularly at high levels and elicited signaling events.

7.2 Homology Modeling

7.2.1 Introduction

The pharmaceutical industry relies heavily on high-throughput screening (HTS) in order to identify potential lead compounds for drug discovery. This method remains expensive, labor and time intensive and overall inefficient.[31] Meanwhile, *in silico* techniques have been incorporated to improve the identification of potential hits. One major drawback of *in silico* methods is that they typically require knowledge of the three-dimensional structure of the target protein. While the number of known protein structures has been increasing dramatically owing to the advances in crystallography, the structural data available are still limited in comparison with the total of known sequences of potential target proteins. In this regard, homology modeling provides a means to predict a three-dimensional model in the absence of any direct structural data and has become a key component in drug design and discovery.

The tertiary structure of a protein is more highly conserved than its amino acid sequence and, therefore, minor changes in the protein's sequence may result in a similar overall protein structure.[32,33] Consequently, the basic principle of homology modeling is that proteins with similar sequences may display common structural features. Homology modeling has successfully generated reliable 3D models in drug discovery and has proven to be a cost-efficient alternative to hit or miss high-throughput screening.[34–37]

On the other hand, there are several steps where homology modeling can go wrong: choosing the wrong template, sequence misalignment and inadequate model refinement. However, in the absence of experimental data, homology modeling remains an important and useful approach in the drug discovery process.

7.2.2   General Procedure

With experimentally established protein structures available to serve as a template, a model of the homologous target protein can be predicted. The procedure to generate a suitable 3D model consists of four major steps: identification of template with known protein structure, sequence alignment of target and template, model generation based off-template, model refinement and validation.[32,38]

The accuracy and reliability of a homology model will only be as good as the sequence homology between the template and target protein. Previous homology modeling studies have shown that sequence identities of 50% or higher typically generate an accurate and reliable model. Models generated from low sequence identities (*e.g.*, below 25%) contain numerous errors and are unreliable.[39] Limited choices of template and sequence misalignments are a major cause of error in homology modeling. Therefore, it is critical to identify a suitable template. Many databases and programs are available for template identification. A simple BLAST or FASTA search can generate a list of potential templates with known protein structures that show a general degree of sequence homology between the target and potential templates.[40] In such a case, where multiple templates are found, there are several considerations to face. Logically, templates carrying the highest homology should be chosen in order to generate the most accurate model. However, several factors must be considered prior to the template selection. Sequence similarity in functionally important regions (e. g. transmembrane helices, catalytic domains and binding pockets) must be heavily weighted in template selection. Ideally, the template should share a high level of homology for those regions since they are of particular interest for drug discovery and design.[37–41] Another important benchmark to consider is the resolution of the template structures. A high-resolution template is desired in order to decrease the likelihood of errors in side-chain placement.[41] Additionally, case-specific considerations may be made based upon factors such as the function and size of the target protein.

Once a template has been identified, a detailed alignment can be performed. On web-based servers such as NCBI, both the template and the target sequence into PSI-BLAST, FASTA or GCG. These programs utilize algorithms with certain parameters such as substitution matrix, gap penalties and significance cutoffs to calculate the best alignment and the degree of similarity.[40] Sometimes, aligning two sequences alone can give misleading results due to low homology, gaps in the alignment and a multitude of other factors. In such cases, using other homologous sequences for a multiple sequence alignment can be beneficial. Specifically, this technique can help identify shared regions of homology among the sequences, predict shared structural features and provide a visual way to quantify the prospected alignments.[37] Programs such as CLUSTALX, PHD or SOPMA can also provide other useful information besides sequence identity. Protein domains, fold structures, residues potentially involved in binding, post-translational modification sites, hydrophobic or hydrophilic areas and secondary structure predictions can be determined from position-specific profiles.[38]

With a template identified, the next step is to generate a model. The model backbone is first built from the residues displaying high sequence homology or the structurally conserved regions (SCR). Since the target and template sequence do not carry 100% homology, deleted and/or inserted sequences will be reflected as gaps between the SCRs in the model.[40] By searching PDB loop structures with end-points that match the amino acid residues at each gap point, the SCRs can be connected.[38] The next step is to incorporate side chains into the structure. Typically, the side chains are created using a library of common rotamers such as SCWRL.

One of final steps is to optimize and verify the model, which may lead back to repeating the previous steps. Programs such as PROCHECK can measure bond lengths and angles and a Ramachandran plot can check torsion angles to verify if they fall within limitations. Visual inspection of the model can also help determine if the structure appears tightly folded by the distribution of polar and non-polar side chains.[38] Experimental data will also be useful in the validation of the model. Site-directed mutagenesis data coupled with binding or protein activity assays can bolster the predicted protein–ligand interactions. For example, the GPCR modeling assessment studies GPCR Dock 2008 and GPCR Dock 2010 both showed that modeling accuracy can be increased by using modeling techniques that are steered with ligand-binding data and available mutagenesis data.[42,43]

7.2.3 G Protein-coupled Receptor Homology Modeling

As of August 2011, there are a total of 357 drugs targeting GPCRs, which account for 36% of all drugs; these formidable numbers serve as an indicator of their importance in both drug discovery and biological systems.[44] A large number of disease states can be attributed directly to the dysfunction of GPCRs and/or their pathways. Of those diseases, cancer has emerged as a particularly useful target for the development of new diagnostic techniques and therapeutics.[45]

Homology modeling's prevalence within GPCR research is caused by the lack of crystal structures for the 390 known GPCRs in the human genome.[46] Owing to the difficulty of crystallizing a transmembrane protein, it was not until 2000 that a high-resolution GPCR crystal structure was available. In 2000, bovine rhodopsin became the first high-resolution GPCR crystal structure that could serve as a suitable template for homology modeling.[47] Consequently, it has been widely used as the template of choice for comparative modeling studies and has helped to explain ligand–receptor interactions, guide mutagenesis studies, reveal structure–activity relationships and perform virtual screenings.[48,49] The next breakthrough in GPCR crystal structures came in 2007 and 2008 with the crystallization of both turkey β_1 and human β_2 adrenergic receptors.[50–52] There were three main techniques that were utilized in the crystallization process: stabilization through an engineered antibody fragment (human β_2), removal and subsequent replacement of the third intracellular loop (IL3) with a readily crystallizable T4-lysozyme (human β_2) and thermostabilization (StaR method, stabilized receptor) through several point mutations (turkey β_1).[50–52] Of these, the thermostabilization and T4-lysozyme methods provided the highest resolutions (2.7 and 2.4 Å, respectively) and revealed that the different crystallization techniques gave comparable structural results.[53] Since then, both methods have been applied to generate structures for the adenosine A_{2A} receptor, chemokine CXCR4 receptor, dopamine D_3 receptor and histamine H1 receptor.[54–58] It is important to note that these structures were co-crystallized with antagonists and only recently have structures for agonist-bound GPCRs been generated using the StaR method. In all, there are seven class A (rhodopsin-like) GPCR crystal structures that have been solved: rhodopsin, β_1 adrenergic receptor, β_2 adrenergic receptor, adenosine A_{2A} receptor, chemokine CXCR4 receptor, dopamine D_3 receptor and histamine H1 receptor.[50–58]

Having an array of different template structures to model GPCRs is a relatively new conundrum. Although the expansion of available crystals structures is sure to provide a new era in GPCR research, it also adds new layers of difficulty in the area of homology modeling. Principally, template selection, which was a non-factor, is now essential. Now multiple sequence alignments must be combined with specific structural and pharmacological knowledge of the template in order to choose the proper template structure.

Several considerations must be taken into account when using these crystal structures as templates. It is important to note that all of the structures contain stabilizing mutations or are chimeric, so they do not fully represent the native structure of any GPCR. Additionally, a GPCR is in constant equilibrium with multiple conformations. However, a model, and especially a crystal structure, is only a still frame of a continuously dynamic system. Therefore, care must be taken when interpreting the data from such a static data source. Information can be distilled from a model, but no model is ever completely correct. That being said, GPCR homology modeling has proven to be a valuable tool for elucidating the structure, function and behavior of these receptors.[59–62]

7.3 G Protein Estrogen Receptor 3D Model and Ligand-binding Mode Study

As discussed above, the construction of a GPCR homology model can be separated into four main steps: identification of template with known protein structure, sequence alignment of target and template, model generation based off-template and model refinement and validation.[32,38] The process is written in a linear fashion, but it has to be performed in a cyclic manner. During every iteration in the process, key decisions must be made to continue or to go back to the previous step in order to achieve the most reliable model possible.

7.3.1 Sequence Alignment

GPCR homology modeling is largely based on the fact that the transmembrane (TM) helices of the receptors share the same relative structure along with several conserved residues. Proper alignment of the target sequence with the template sequence is essential and serves as the basis of any robust model. For this study's purposes, the sequence alignment program BioEdit was used; however, several other programs are also available.[63]

Since all GPCRs have a common seven TM domain motif and several conserved residues, most alignment programs can give adequate initial results

```
2VT4    -----------------------------------------MG----------
2RH1    ------------------------------DYKDDDDAMGQP------~G
3EML    ----------------------------DYKDDDDAMGQP------~V
3RZE    ------------------------------------------------------
3PBL    ----------------------------DYKDDDDGAPAS-------L
3ODU    ----------------------------DYKDDDDAGAPEGISIYTSD
1U19    ------------------------------XMNGTEGPNFYVPFSNKTG
GPER    MDVTSQARGVGLEMYPGTAQPAAPNTTSPELNLSHPLLGTALANGTGELS


2VT4    ---AELLS--------------QQWEAGMSLLMALVVLLIVAGNVLVI
2RH1    NGSAFLLAPNRSHAPDHDVTQQRDEVWVVGMGIVMSLIVLAIVFGNVLVI
3EML    GAPPIMGS-----------------SVYITVELAIAVLAILG-NVLVC
3RZE    ---TTMAS-----------------PQLMPLVVVLSTICLVTVGLNLLVL
3PBL    SQLSSHLNYTCG---AENSTGASQARPHAYYALSYCALILAIVFGNGLVC
3ODU    NYTEEMGSGDYDSMKEPCFREENANFNKIFLPTIYSIIFLTGIVGNGLVI
1U19    VVRSPFEAP--------QYYLAEPWQFSMLAAYMFLLIMLGFPINFLTL
GPER    EHQQYVIG-------------------LFLSCLYTIFLFPIGFVGNILIL
                                        .    *  *


2VT4    AAIGSTQRLQTLTNLFITSLACADLVVGLLVVPFGATLVVRG-TWLWGSF
2RH1    TAIAKFERLQTVTNYFITSLACADLVMGLAVVPFGAAHILMK-MWTFGNF
3EML    WAVWLNSNLQNVTNYFVVSLAAADIAVGVLAIPFAITISTG---FCAACH
3RZE    YAVRSERKLHTVGNLYIVSLSVADLIVGAVVMPMNILYLLMS-KWSLGRP
3PBL    MAVLKERALQTTTNYLVVSLAVADLLVATLVMPWVVYLEVTGGVWNFSRI
3ODU    LVMGYQKKLRSMTDKYRLHLSVADLLF-VITLPFWAVDAVAN--WYFGNF
1U19    YVTVQHKKLRTPLNYILLNLAVADLFMVFGGFTTTLYTSLHG-YFVFGPT
GPER    VVNISFREKMTIPDLYFINLAVADLILVADSLIEVFNLHER---YYDIAV
          .          .   :      *:  **:  .      .              :
```

Figure 7.3 Multiple sequence alignment of GPER and seven template structure sequences.

```
2VT4      LCELWTSLDVLCVTASIETLCVIAIDRYLAITSPFRYQS---LMTRARAK
2RH1      WCEFWTSIDVLCVTASIETLCVIAVDRYFAITSPFKYQS---LLTKNKAR
3EML      GCLFIACFVLVLTQSSIFSLLAIAIDRYIAIRIPLRYNG---LVTGTRAK
3RZE      LCLFWLSMDYVASTASIFSVFILCIDRYRSVQQPLRYLK---YRTKTRAS
3PBL      CCDVFVTLDVMMCTASIWNLCAISIDRYTAVVMPVHYQHGTG-QSSCRRV
3ODU      LCKAVHVIYTVNLYSSVWILAFISLDRYLAIVHATNSQR----PRKLLAE
1U19      GCNLEGFFATLGGEIALWSLVVLAIERYVVVCKPMSNFR----FGENHAI
GPER      LCTFMSLFLQVNMYSSVFFLTWMSFDRYIALARAMRCSL---FRTKHHAR
           *       :    :     ::   :   :..:**  :   .


2VT4      VIICTVWAISALVSFLPIMMHWWRDED-PQALKCYQDPGCCDFVTNRAYA
2RH1      VIILMVWIVSGLTSFLPIQMHWYRATH-QEAINCYAEETCCDFFTNQAYA
3EML      GIIAICWVLSFAIGLTPMLGWNNCGQP-KEGKNHSQGCGEGQVACLFEDV
3RZE      ATILGAWFLS-FLWVIPILGWNHFMQQ-TSVR--REDKCETDFYDVTWFK
3PBL      ALMITAVWVLAFAVSCPLLFGFNTTGD-PTVCS---------ISNPDFV
3ODU      KVVYVGVWIPALLLTIPDFIFANVSEAD-DRYICDR------FYPNDLWV
1U19      MGVAFTWVMALACAAPPLVGWSRYIPEGMQCSCGIDYYTPHEETNNESFV
GPER      LSCGLIWMASVSATLVPFTAVHLQHTD-EACFC---------FADVREVQ
                                  *


2VT4      IASSIISFY-------IPLLIMIFVALRVYREAKEQIRKIDRASKRKRVM
2RH1      IASSIVSFY-------VPLVIMVFVYSRVFQEAKRQLK-----------
3EML      VPMNYMVYFNFFACVLVPLLLMLGVYLRIFLAARRQLR---------ST
3RZE      VMTAIINFY-------LPTLLMLWFYAKIYKAVRQHCLH----------
3PBL      IYSSVVSFY-------LPFGVTVLVYARIYVVLKQRRRK---------G
3ODU      VVFQFQHIMVG---LILPGIVILSCYCIIISKLSHSGSG--------SK
1U19      IYMFVVHFI-------IPLIVIFFCYGQLVFTVKEAAAQQQES----ATT
GPER      WLEVTLGFI-------VPFAIIGLCYSLIVRVLVRAHRHRG-------L
                  :*  :         :         .


2VT4      LMREHKALKTLGIIMGVFTLCWLPFFL------VNIVNVFNRDLVPDWLF
2RH1      FLKEHKALKTLGIIMGTFTLCWLPFFI------VNIVHVIQDNLIRKEVY
3EML      LQKEVHAAKSLAIIVGLFALCWLPLHI------INCFTFFCPDCSHAPLW
3RZE      MNRERKAAKQLGFIMAAFILCWIPYFI------FFMVIAFCKNCCNEHLH
3PBL      VPLREKATQMVAIVLGAFIVCWLPFFL------THVLNTHCQTCHVSPEL
3ODU      GHQKRKALKTTVILILAFFACWLPYYIGISIDSFILLEIIKQGCEFENTV
1U19      QKAEKEVTRMVIIMVIAFLICWLPYAG------VAFYIFTHQGSDFGPIF
GPER      RPRRQKALRMILAVVLVFFVCWLPENVFIS---VHLLQRTQPGAAPCKQS
              .  ..:     ::   *   **:*
```

Figure 7.3   Continued

```
2VT4      VAFN--------WLGYANSAMNPIIYCRSP-DFRKAFKRLLAFPRKADR
2RH1      ILLN--------WIGYVNSGFNPLIYCRSP-DFRIAFQELLCLRRSSLK
3EML      LMYLAI------VLSHTNSVVNPFIYAYRIREFRQTFRKIIRSHVLRQQ
3RZE      MFTI--------WLGYINSTLNPLIYPLCNENFKKTFKRILHIRSGENL
3PBL      YSATT-------WLGYVNSALNPVIYTTFNIEFRKAFLKILSCGRPLEV
3ODU      HKWISITE-----ALAFFHCCLNPILYAFLGAKFKTSAQHALTSGRPLEV
1U19      MTIP--------AFFAKTSAVYNPVIYIMMNKQFRNCMVTTLCCGKNPLG
GPER      FRHAHPLTGHIVNLAAFSNSCLNPLIYSFLGETFRDKLRLYIEQKTNLPA
                     .    .  **.:*        *:          :


2VT4      RLHHHHHH------------------
2RH1      AYGNGYSSNGNTGEQSG-----------
3EML      EPFKAHHHHHHHHHH-------------
3RZE      YFQ-----------------------
3PBL      LFQ-----------------------
3ODU      LFQ-----------------------
1U19      DDEASTTVSKTETSQVAPA---------
GPER      LNRFCHAALKAVIPDSTEQSDVRFSSAV
```

Figure 7.3 Continued

to distinguish the individual templates from each other. Once a template has been selected, a more comprehensive sequence alignment must be performed. Initially, GPER was aligned with seven high-resolution GPCR crystal structures: turkey β_1 adrenergic receptor, human β_2 adrenergic receptor, human A_{2A} adenosine receptor, human histidine H_1 receptor, human CXCR4 chemokine receptor and bovine rhodopsin (PDB code: 2VT4, 2RH1, 3EML, 3RZE, 3PBL, 3ODU and 1U19, respectively) (Figure 7.3). As expected, the highly conserved residues Asn1.5, Leu2.5, Arg3.5, Trp4.5, Pro5.5, Pro6.5 and Pro7.5 all aligned properly amongst the sequences and there were no significant gaps in the transmembrane helical domains.

7.3.2   Template Selection

Several factors weigh in when considering which template and subsequent alignment to use. The overall homology is important, of course, but it can be more critical to look at the homology levels of the individual helices. Variation among the intra- and extracellular loops is expected, but the helical bundles share more residues between each other and therefore serve as a better judgment for homology. Table 7.1 illustrates the homology between GPER and the individual template structure regions.

2VT4 (turkey β_1 adrenergic) was automatically eliminated from the possible template structures owing its overall low homology. Besides, it was crystallized

Table 7.1 GPER homology between template structures in individual helices (TM1-7), intra- (IL1-3) and extracellular (EL1-3) loops. Numbers shown are ratios of identity + similarity/total number of residues in that region. Bold italic numbers represent the highest homology in that section.

| | *GPCR template* | | | | | |
	1U19	*2RH1*	*3PBL*	*3ODU*	*3EML*	*3RZE*
N-terminus	0.23	0.28	0.29	0.24	0.17	0.00
TM1	0.5	0.51	0.39	*0.65*	0.45	0.30
IL1	0.29	0.33	0.25	0.4	0.2	0.33
TM2	*0.57*	0.57	0.47	0.57	0.48	0.53
EL1	0	0	1	0.25	0.33	0.20
TM3	0.49	0.54	0.63	*0.77*	0.57	0.41
IL2	0.38	0.4	0.3	0.33	0.38	0.57
TM4	0.54	*0.68*	0.39	0.33	0.46	0.24
El2	0.2	0.44	0.07	0.46	0.4	0.00
TM5	0.44	0.42	0.4	*0.53*	0.48	0.43
IL3	0.41	0.13	0.43	0.2	0.67	0.41
TM6	0.56	0.62	*0.83*	0.77	0.49	0.56
EL3	0.17	0.29	0.32	0	0.29	0.17
TM7	0.52	0.52	0.44	0.47	*0.56*	0.52
C-terminus	0.29	0.31	0.15	0.22	0.26	0.29
Average of TMs	0.52	0.55	0.51	0.58	0.50	0.43

in a thermostabilized approach with multiple point mutations to lock the receptor in a single conformation. Although this aided in the crystallization process, it may be less representative of the receptor in its natural state and environment. The other representative template structures were crystallized by inserting a T4 lysozyme into the third intracellular loop, which does not change the receptor's thermal stability, but only aids in its crystallization. Another drawback of the thermal stabilization method (StaR, 'stabilized receptor') is that the point mutations present may alter the receptor's affinity for its ligands. Each method has pros and cons but overall the T4 lysozyme insertion method may represent a more 'native' model since it may not directly alter the thermal stabilization or ligand-binding mode of the receptor.[53]

The results in Table 7.1 suggest that 3ODU (human CXCR4 chemokine) shares the highest homology within the helical regions to GPER; it also has several helices with very high homology to GPER. Again, homology is not the only factor when choosing a template structure. Since 3ODU is a chemokine receptor, its natural ligands are large peptides which are significantly different from the small, rigid ligands of GPER. Additionally, there are several structural features seen in 3ODU that are not common in the aminergic receptors. For example, several helices are shifted, rotated, or even longer than the other crystal structures.[53] For these reasons, 3ODU may not be an ideal template structure for GPER.

Behind 3ODU, 2RH1 (human β_2 adrenergic) has the second highest homology to GPER in its transmembrane helices and also similar homology to the rest of the receptor. Moreover, as an adrenergic receptor, 2RH1 has small-molecule ligands, which are more similar to GPER's. Looking further at the sequence alignment between 2RH1 and GPER, there are no gaps in the helices' alignment and there is also high homology in extracellular loop 2 (EL2). The involvement of extracellular loop 2 in GPCR ligand binding is well known and, therefore, extra care should be taken at this stage of the modeling process.

Overall, compared with all other possible template structures, 2RH1 appears to be the best choice of the available crystal structures. An alternative method to using a single template structure to model GPER would be to use multiple templates to model it. The method uses an amalgam of alignments from different template structures having the highest homology and models the protein by using specified sections of each. Although this method is more time consuming and complicated (not being used in this study), it can provide valuable structural information when a single template fails to give reasonable results.[37]

7.3.3   Model Generation, Selection and Optimization

Before generating any models, the alignment between 2RH1 and GPER was optimized. During this process, several factors were considered: gaps in helix alignments, position of conserved residues and position of key cysteine residues. Figure 7.4 shows the optimized alignment that was applied to produce prospective GPER models. An important aspect to note is that both the N- and C-termini of GPER were shortened, by 49 and 53 residues, respectively, because there is no template to model them because of their length and flexibility.

Upon final optimization of the pairwise alignment of GPER and 2RH1, homology models of GPER were produced by using the modeling program Modeller 9v8.[64] Modeller works by using a satisfaction of spatial restraints in order to map the location of each atom based on the template structure and the sequence alignment. A total of 100 models were generated from this process.

Since multiple solutions are generated from Modeller, a selection has to be made. Along with the models, three assessment scores (molpdf, DOPE and GA341) were generated to indicate the general fitness and 'nativeness' of the receptor. They served to show how well each model ranked among the group, but no information was gained as how these receptor models would interact with their ligands. In this regard, a pertinent way to make a selection is to dock each model with some known ligands. From there, the differences between the models can be quantified based on how well they bind with the ligands. For this purpose, the GPER-selective antagonists G-15 and G-36 were docked into each receptor using GOLD 5.1.[65]

Ligand-docking techniques (*e.g.*, GOLD, AutoDock) can be used to place the compounds in the binding sites of the receptor models in a favorable

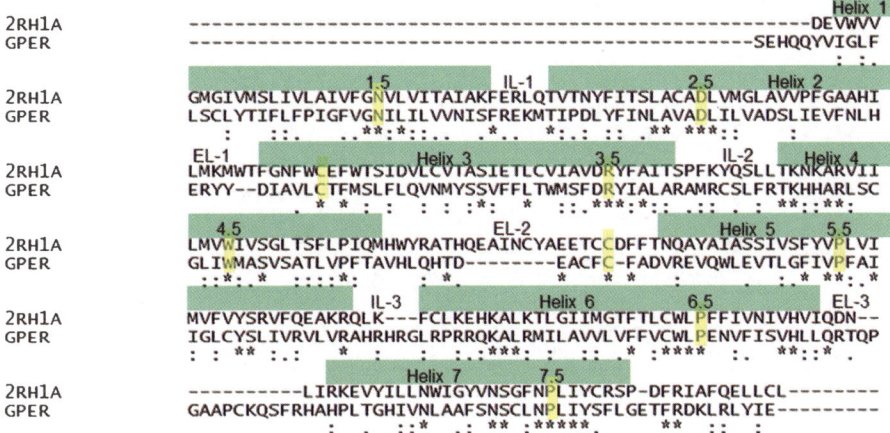

Figure 7.4 Optimized alignment between 2RH1 and GPER. Residues highlighted in yellow are conserved in all GPCRs.

manner. Docking-mode differences between the models are quantitated by using the GOLD scores from the docking. Overall, this process is more subjective than it sounds. Models and bound ligands have to be sieved through to elucidate which binding mode is the most reasonable based on the reported data available. The model picked out from the process bound G-15 with the highest GOLD score and G-36 with a relatively high GOLD score. Furthermore, when analyzed using a Ramachandran analysis (Figure 7.5), there were no outliers within the transmembrane helices; only three outliers within the loops were observed. These outliers can be fixed through further optimization.

The prospective homology model selected from ligand-docking studies has to be energy minimized and checked for any errors. The GPER model refinement was carried out through energy minimization of the receptor–ligand complex using Sybyl 8.1. Several energy minimization force fields can be used in the optimization, such as Tripos, MMFF94 (Merck molecular force field) and AMBER (assisted model building with energy refinement). All three force fields have been validated and shown to work on energy refinement for large biomolecules.[66–69] There are several fundamental differences between each force field in how they use and apply physical approximations in order to minimize the energy of a macromolecule.[67] Therefore, all three of them were applied to see which would give the best results. After minimizing the complex with all three force fields, the model quality was assessed by using PROTABLE and PROCHECK.[70] Both the Tripos and AMBER force fields failed to resolve any outliers in the Ramachandran analysis. However, the MMFF94 force field was able to resolve all but one outlier and gave an overall lower energy confirmation of the receptor. Figure 7.6 shows the minimized GPER model docked with G-15.

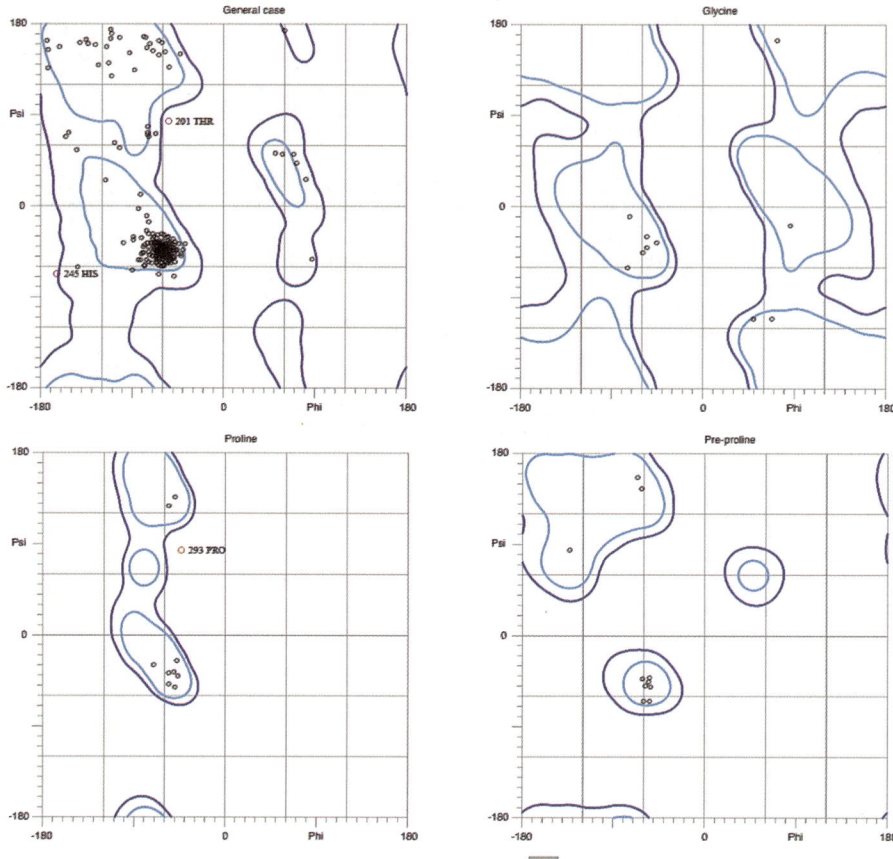

Figure 7.5   Ramachandran analysis of GPER model before loop remodeling.

In order to test the model's trustworthiness further, the selective antagonists were docked into the chosen homology model and subsequently superimposed within the binding pocket (Figure 7.7). The docked solutions of the compounds can be combined with their associated potency, affinity and selectivity data to validate the modeled receptor–ligand complexes. The modeled complexes must be consistent with the experimental data; if not, the model must be modified, refined, or abandoned. An additional consideration is that a limitation of docking algorithms, such as GOLD, is their inability to account for the inherent flexibility of the side chains in the receptor, which can skew the docking solutions. There are multiple ways in which this shortcoming can be addressed: flexible side chains can be explicitly incorporated in the docking algorithm (*i.e.*, conformers can be sampled from a rotamer library), several conformations of the receptor can be used for the docking and side chains can be manually adjusted (within the realm of plausibility). All of these techniques can be utilized with the aim of improving the model's agreement with experimental data.

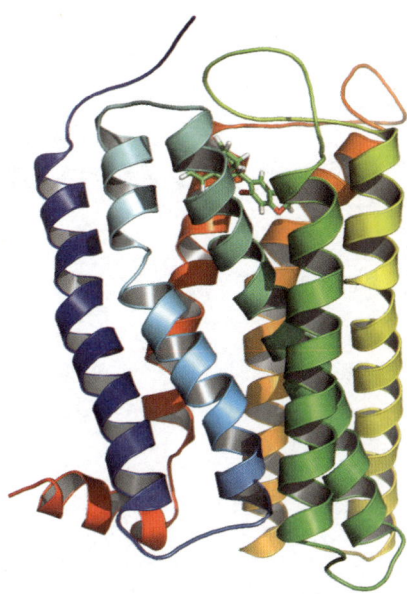

Figure 7.6   GPER homology model with G-15 bound.

Several favorable interactions were observed between GPER and the antagonists (Figure 7.8). Asn310 may act as a hydrogen bond acceptor, interacting with the secondary amino group of the antagonists' tetrahydroquinoline moiety. There are several hydrophobic interactions that occur along the backbone of the molecules with Pro303, Phe278, Ile279, Phe206, Leu137, Met133, Val116 and Leu119. Additionally, Phe208 located in the EL2 is involved with pi–pi stacking and hydrophobic interactions with the 1,3-benzodioxole ring of G-15 and G-36. These interactions were seen for both G-15 and G-36, which suggests a common pharmacophore model for antagonists.

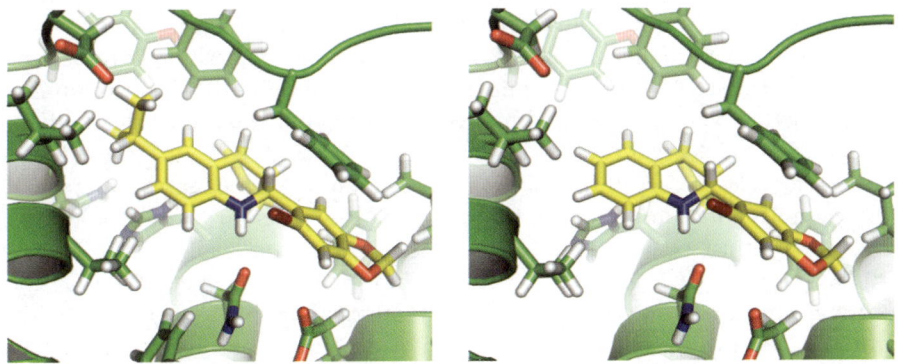

Figure 7.7   G-15 and G-36 binding to the GPER homology model.

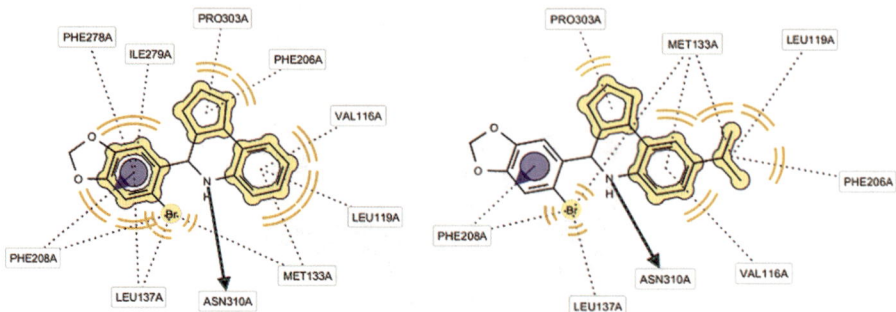

Figure 7.8   Possible interactions between GPER and selective ligands G-15 and G-36.

7.4   Discussion

No model is ever completely accurate; it can always be reanalyzed and optimized. Ideally, there would be more data to scrutinize the GPER model further, but currently there is only one set of ligands with very limited structure–activity relationship data (G-1, G15 and G-36). Additionally, the GPER model needs to be verified with site-directed mutagenesis data or cysteine-scanning mutagenesis. Doing so would indicate which residues within the receptor are important for ligand binding and/or function.

Further work could be carried out to optimize intra- and extracellular loops so as to make a more accurate model, since these structures typically have little similarity to the identity and length of the template sequence. Extracellular loop 2 is of special importance since it has been shown to play a role in ligand recognition and binding.[53,62] Several methods, such as MODELLER's loop refinement and CHARMm, can be employed to refine loops. Molecular dynamics simulations can also be performed to equilibrate the receptors in an aqueous membrane system to refine the loops between the transmembrane helices.

The information gained from the GPER homology model can be applied to several structural and pharmacological aspects of GPCR research. One of the most pertinent uses of the homology model is new ligand design. A general pattern was seen for the binding of G-15 and G-36, which suggests a number of key residues within the binding pocket. A putative pharmacophore can be suggested (Figure 7.9) and this information can be applied to lead optimization, structure–activity relationship rationalization and a better understanding of the pharmacological profile of the receptor.

Another relevant use of the GPER homology model is to identify amino acid residues relevant to its biological function.[37] Since models are only an educated guess, they must be backed up with experimental data such as mutagenesis studies. However, before such studies can be performed, a general idea of the protein structure and receptor–ligand interaction has to be deduced. Homology modeling can aid in the design of the studies by suggesting residues of interest. In turn, the mutagenesis data can be applied to modify the model to make it more accurate.

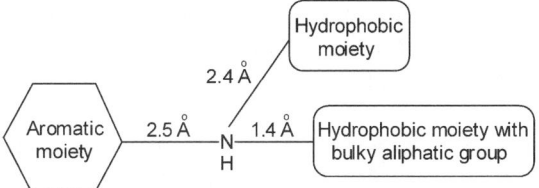

Figure 7.9   Antagonist pharmacophore based on the binding of G-15 and G-36 to GPER homology model. Distances reported are distances from an amine capable of hydrogen bond donation.

The knowledge gained from the GPER homology modeling studies can be directly applied to other nuclear GPCRs. The methodology developed can be followed to model, screen and understand other nuclear GPCRs. Furthermore, by increasing the amount of modeled nuclear GPCRs, common features among them may be of help in understanding better their utility and function. An important aspect of the methodology developed is the special attention paid to template choice and alignment to the target sequence. Within GPCR homology modeling, the number of template structures has increased dramatically and will continue to do so. The seven templates utilized in this study will most likely represent just a fraction of GPCR template structures in the years to come. Therefore, choosing a template structure from the myriad of possible choices will become increasingly difficult. The methodology applied in GPER homology modeling provides a logical, stepwise route to help in the decision and thus aid in the modeling of other nuclear GPCRs.

In summary, the GPER model can provide insight into both methodology development and ligand design. In relation to other nuclear GPCRs, the modeling procedure utilized for GPER provides a stepwise and logical methodology. In addition to the methodology developed, the GPER homology model also provides important information about ligand binding. Homology modeling provides a fast and economical route to gain a multitude of information about a GPCR of interest.

References

1. C. J. Gruber, W. Tschugguel, C. Schneeberger and J. C. Huber, *N. Engl. J. Med.*, 2002, **346**, 340.
2. E. R. Prossnitz, J. B. Arterburn, H. O. Smith, T. I. Oprea, L. A. Sklar and H. J. Hathaway, *Annu. Rev. Physiol.*, 2008, **70**, 165.
3. C. Carmeci, D. A. Thompson, H. Z. Ring, U. Francke and R. J. Weigel, *Genomics*, 1997, **45**, 607.
4. E. J. Filardo, J. A. Quinn, K. I. Bland and A. R. Frackelton Jr, *Mol. Endocrinol.*, 2000, **14**, 1649.
5. E. R. Prossnitz, J. B. Arterburn and L. A. Sklar, *Mol. Cell Endocrinol.*, 2007, **256**, 138.

6. E. R. Prossnitz, L. A. Sklar, T. I. Oprea and J. B. Arterburn, *Trends Pharm. Sci.*, 2008, **29**, 116.
7. E. R. Prossnitz and M. Barton, *Prostaglandins Other Lipid Mediat.*, 2009, **89**, 89.
8. E. R. Prossnitz and M. Barton, *Nat. Rev. Endocrinol.*, 2011, **7**, 175.
9. E. J. Filardo, C. T. Graeber, J. A. Quinn, M. B. Resnick, D. Giri, R. A. DeLellis, M. M. Steinhoff and E. Sabo, *Clin. Cancer Res.*, 2006, **12**, 6359.
10. D. Wang, L. Hu, G. Zhang, L. Zhang and C. Chen, *Endocrinology*, 2010, **38**, 29.
11. H. Arias-Pulido, M. Royce, Y. Gong, N. Joste, L. Lomo, S.-J. Lee, N. Chaher, C. Verschraegen, J. Lara, E. R. Prossnitz and M. Cristofanilli, *Breast Cancer Res. Treat.*, 2010, **123**, 51.
12. H.-J. Luo, P. Luo, G.-L. Yang, Q.-L. Peng, M.-R. Liu and G. Tu, *J. Breast Cancer*, 2011, **14**, 185.
13. A. Ignatov, T. Ignatov, A. Roessner, S. D. Costa and T. Kalinski, *Breast Cancer Res. Treat.*, 2010, **123**, 87.
14. C. G. Bologa, C. M. Revankar, S. M. Young, B. S. Edwards, J. B. Arterburn, A. S. Kiselyov, M. A. Parker, S. E. Tkachenko, N. P. Savchuck, L. A. Sklar, T. I. Oprea and E. R. Prossnitz, *Nat. Chem. Biol.*, 2006, **2**, 207.
15. M. K. Dennis, R. Burai, C. Ramesh, W. K. Petrie, S. N. Alcon, T. K. Nayak, C. G. Bologa, A. Leitao, E. Brailoiu, E. Deliu, N. J. Dun, L. A. Sklar, H. J. Hathaway, J. B. Arterburn, T. I. Oprea and E. R. Prossnitz, *Nat. Chem. Biol.*, 2009, **5**, 421.
16. M. K. Dennis, A. S. Field, R. Burai, C. Ramesh, W. K. Petrie, C. G. Bologa, T. I Oprea, Y. Yamaguchi, S.-I. Hayashi, L. A. Sklar, H. J. Hathaway, J. B. Arterburn and E. R. Prossnitz, *J. Steroid Biochem. Mol. Biol.*, 2011, **127**, 358.
17. F. Gobeil, A. Fortier, T. Zhu, M. Bossolasco, M. Leduc, M. Grandbois, N. Heveker, G. Bkaily, S. Cehmtob and D. Barbaz, *Can. J. Physiol. Pharmacol.*, 2006, **84**, 287.
18. D. Lu, H. Yang, G. Shaw and M. K. Raizada, *Endocrinology*, 1998, **139**, 365.
19. R. Chen, Y. V. Mukhin, M. N. Garnovskaya, T. E. Thielen, Y. Iijima, C. Huang, J. R. Raymond, M. E. Ullian and R. V. Pual, *Am. J. Physiol. Renal Physiol.*, 2000, **279**, 440.
20. D. K. Lee, A. J. Lanca, R. Cheng, T. Nguyen, X. D. Ji, F. Gobeil Jr, S. Chemtob, S. R. George and B. F. O'Dowd, *J. Biol. Chem.*, 2004, **279**, 7901.
21. A. Tadevosyan, A. Maguy, L. R. Villeneuve, J. Babin, A. Bonnefoy, B. G. Allen and S. Nattel, *J. Biol. Chem.*, 2010, **285**, 22338.
22. F. Gobeil Jr, I. Dumont, A. M. Marrache, A. Vazquez-Tello, S. G. Bernier, D. Abran, X. Hou, M. H. Beauchamp, C. Quiniou, A. Bouayad, S. Choufani, M. Bhattacharya, S. Molotchnikoff, A. Ribeiro-Da-Silva, D. R. Varma, G. Bkaily and S. Chemtob, *Circ. Res.*, 2002, **90**, 682.
23. K. L. O'Malley, Y. J. Jong, Y. Gonchar, A. Burkhalter and C. Romano, *J. Biol. Chem.*, 2003, **278**, 28210.

24. B. Boivin, D. Chevalier, L. R. Villeneuve, E. Rousseau and B. G. Allen, *J. Biol. Chem.*, 2003, **278**, 29153.

25. F. Gobeil Jr, S. G. Bernier, A. Vazquez-Tello, S. Brault, M. H. Beauchamp, C. Quiniou, A. M. Marrache, D. Checchin, F. Sennlaub, X. Hou, M. Nader, G. Bkaily, A. Ribeiro-da-Silva, E. J. Goetzl and and S. Chemtob, *J. Biol. Chem.*, 2003, **278**, 38875.

26. C. D. Wright, Q. Chen, N. L. Baye, Y. Huang, C. L. Healy, S. Kasinathan and T. D. O'Connell, *Circ. Res.*, 2008, **103**, 992.

27. A. M. Marrache, F. Gobeil Jr, S. G. Bernier, J. Stankova, M. Rola-Pleszczynski, S. Choufani, G. Bkaily, A. Bourdeau, M. G. Sirois, A. Vazquez-Tello, L. Fan, J. S. Joyal, J. G. Filep, D. R. Varma, A. Ribeiro-Da-Silva and S. Chemtob, *J. Immunol.*, 2002, **169**, 6474.

28. G. Vaniotis, B. G. Allen and T. E. Hébert, *Am. J. Physiol. Heart Circ. Physiol.*, 2011, **301**, 1754.

29. C. M. Revankar, D. F. Cimino, L. A. Sklar, J. B. Arterbum and E. R. Prossnitz, *Science*, 2005, **307**, 1625.

30. C. M. Revankar, H. D. Mitchell, A. S. Field, R. Burai, C. Corona, C. Ramesh, L. A. Sklar, J. B. Arterburn and E. R. Prossnitz, *ACS Chem. Biol.*, 2007, **2**, 536.

31. M. Congreve, C. W. Murray and T. L. Blundell, *Drug Discov. Today*, 2005, **10**, 895.

32. A. Hillisch, L. F. Pinead and R. Hilgenfeld, *Drug Discov. Today*, 2004, **9**, 659.

33. A. M. Lesk and C. H. Chothia, *Philos. Trans. R. Soc. Lond. B Biol. Sci.*, 1986, **317**, 345.

34. I. T. Weber, *Proteins*, 1990, **7**, 172.

35. A. Hillisch, L. F. Pineda and R. Hilgenfeld, *Drug Discov. Today*, 2004, **9**, 659.

36. I. Sela, G. Golan, M. Strajbl, D. Rivenzon-Segal, S. Bar-Haim, I. Bloch, B. Inbal, A. Shitrit, E. Ben-Zeev, M. Fichman, Y. Markus, Y. Marantz, H. Senderowitz and O. Kalid, *Curr. Top. Med. Chem.*, 2010, **10**, 638.

37. C. N. Cavasotto and S. S. Phatak, *Drug Discov. Today*, 2009, **14**, 676.

38. E. Krieger, S. B. Nabuurs and G. Vriend, in *Structural Bioinformatics*, ed. P. E. Bourne and H. Weissig, Wiley-Liss, Hoboken, NJ, 2003, p. 509.

39. M. J. Forster, *Micron*, 2002, **33**, 365.

40. A. Tramontano, *Methods*, 1998, **14**, 293.

41. J. C. Mobarec, R. Sanchez and M. Filizola, *J. Med. Chem.*, 2009, **52**, 5207.

42. M. Michino, E. Abola, GPCR Dock 2008 Participants, C. L. Brooks III, J. S. Dixon, J. Moult and R. C. Stevens, *Nat. Rev. Drug Discov.*, 2009, **8**, 455.

43. I. Kufareva, M. Rueda, V. Katritch, GPCR Dock 2010 Participants, R. C. Stevens and R. Abagyan, *Structure*, 2011, **19**, 1108.

44. M. Rask-Andersen, M. S. Almén and H. B Schiöth, *Nat. Rev. Drug Discov.*, 2011, **10**, 579.

45. R. Lappano and M. Maggiolini, *Nat. Rev. Drug Discov.*, 2011, **10**, 47.

46. M. C. Lagerstrom and H. B. Schioth, *Nat. Rev. Drug Discov.*, 2002, **1**, 727.

47. K. Palczewski, T. Kumasaka, T. Hori, C. A. Behnke, H. Motoshima, B. A. Fox, I. Le Trong, D. C. Teller, T. Okada, R. E. Stenkamp, M. Yamamoto and M. Miyano, *Science*, 2000, **289**, 739.
48. A. Patny, P. V. Desai and M. A. Avery, *Curr. Med. Chem.*, 2006, **13**, 1667.
49. C. Bizzantz, P. Bernard, M. Hibert and D. Rognan, *Proteins: Struct. Funct. Bioinf.*, 2003, **50**, 5.
50. S. G. F. Rasmussen, H.-J. Choi, D. M. Rosenbaum, T. S. Kobilka, F. S. Thian, P. C. Edwards, M. Burghammer, V. R. P. Ratnala, R. Sanishvili, R. F. Fischetti, G. F. X. Schertler, W. I. Weis and B. K. Kobilka, *Nature*, 2007, **450**, 383.
51. V. Cherezov, D. M. Rosenbaum, M. A. Hanson, S. G. Rasmussen, F. S. Thian, T. S. Kobilka, H. J. Choi, P. Kuhn, W. I. Weis, B. K. Kobilka and R. C. Stevens, *Science*, 2007, **318**, 1258.
52. T. Warne, M. J. Serrano-Vega, J. G. Baker, R. Moukhametzianov, P. C. Edwards, R. Henderson, A. G. Leslie, C. G. Tate and G. F. Schertler, *Nature*, 2008, **454**, 486.
53. M. Congreve, C. J. Langmead, J. S. Mason and F. H. Marshal, *J. Med. Chem.*, 2011, **54**, 4283.
54. T. Okada, M. Sugihara, A. N. Bondar, M. Elstner, P. Entel and V. Buss, *J. Mol. Biol.*, 2004, **342**, 571.
55. V. P. Jaakola, M. T. Griffith, M. A. Hanson, V. Cherezov, E. Y. Chien, J. R. Lane, A. P. Ijzerman and R. C. Stevens, *Science*, 2008, **322**, 1211.
56. B. Wu, E. Y. Chien, C. D. Mol, G. Fenalti, W. Liu, V. Katritch, R. Abagyan, A. Brooun, P. Wells, F. C. Bi, D. J. Hamel, P. Kuhn, T. M. Handel, V. Cherezov and R. C. Stevens, *Science*, 2010, **330**, 1066.
57. E. Y. Chien, W. Liu, Q. Zhao, V. Katritch, G. W. Han, M. A. Hanson, L. Shi, A. H. Newman, J. A. Javitch, V. Cherezov and R. C. Stevens, *Science*, 2010, **330**, 1091.
58. T. Shimamura, M. Shiroishi, S. Weyand, H. Tsujimoto, G. Winter, V. Katritch, R. Abagyan, V. Cherezov, W. Liu, G. W. Han, T. Kobayashi, R. C. Stevens and S. Iwata, *Nature*, 2011, **475**, 65.
59. K. K. Roy and A. K. Saxena, *J. Chem. Inf. Model.*, 2011, **51**, 1405.
60. V. Katritch, M. Rueda, P. C.-H. Lam, M. Yeager and R. Abagyan, *Proteins*, 2009, **78**, 197.
61. J. Carlsson, R. G. Coleman, V. Setola, J. J. Irwin, H. Fan, A. Schlessinger, A. Sali, B. L. Roth and B. K. Shoichet, *Nat. Chem. Biol.*, 2011, **7**, 769.
62. F. Fanelli and P. G. De Benedetti, *Chem. Rev.*, 2011, **111**, PR43.
63. T. A. Hall, *Nucleic Acids Symp. Ser.*, 1999, **41**, 95.
64. A. Sali and T. L. Blundell, *J. Mol. Biol.*, 1993, **234**, 779.
65. M. L. Verdonk, J. C. Cole, M. J. Hartshorn, C. W. Murray and R. D. Taylor, *Proteins*, 2003, **52**, 609.
66. M. Clark, R.D. Cramer III and N. Van Opdenbosch, *J. Comput. Chem.*, 1989, **10**, 982.
67. T. Halgren, *J. Comput. Chem.*, 1996, **17**, 720.
68. T. Halgren, *J. Comput. Chem.*, 1999, **20**, 720.

69. W.D. Cornell, P. Cieplak, C. I. Bayly, I. R. Gould, K. M. Merz Jr, D. M. Ferguson, D. C. Spellmeyer, T. Xov, J. W. Caldwell and P. A. Kollman, *J. Am. Chem. Soc.*, 1995, **117**, 5179.
70. R. A. Laskowski, M. W. Macarthur, D. S. Moss and J. M. Thornton, *J. Appl. Crystallogr.*, 1993, **26**, 283.

CHAPTER 8

Reporter Bioluminescent Mice to Test Computational Studies

SARA DELLA TORRE AND ADRIANA MAGGI*

Center of Excellence on Neurodegenerative Diseases and Department of
Pharmacological Sciences, University of Milan, via Balzaretti 9, 20133 Milan, Italy
*E-mail: adriana.maggi@unimi.it

8.1   Introduction

Estrogen receptors (ERs) belong to a large family of molecules of growing interest because of their relevance for the control of reproductive and major metabolic functions in mammals.[1] The two ERs described in mammals (ERα and ERβ) are present in a wide variety of cells and tissues and are considered as ubiquitously expressed. ERs are maintained inactive by a series of binding proteins masking their functional domains; once activated by their cognate ligands, ERs undergo a series of structural modifications enabling the receptor protein to interact with other macromolecules present both in the cytoplasm and in the nucleus. In fact, a large body of evidence has demonstrated that ERs may regulate cell metabolism either by directly binding specific DNA sequences [named estrogen-responsive elements (EREs)] and regulating gene transcription or by binding to other signaling molecules and interfering with their activities. The consequence of such a complex mechanism of action is that ERs may modulate a large number of intracellular processes of pathophysiological relevance.

Most interestingly, the intracellular activity of ERs may be triggered by a large variety of molecules, including growth factors, neurotransmitters and a variety of steroidal and non-steroidal compounds.

RSC Drug Discovery Series No. 30
Computational Approaches to Nuclear Receptors
Edited by Pietro Cozzini and Glen E. Kellogg
© The Royal Society of Chemistry 2012
Published by the Royal Society of Chemistry, www.rsc.org

The complexity of the mechanism of action and activities of ERs together with the observation that these proteins are the most ancient of all intracellular receptors suggests that originally these molecules might have been operating as general transcription factors; then, with the evolution and perfection of the estrogen biosynthetic pathway, ERs acquired more specialized functions.[2,3] As we might expect on the basis of the above, the pharmacology of these receptors is extremely complex as most of estrogenic compounds (steroidal or non-steroidal) exhibit an activity on the receptor that is tissue specific. Therefore, molecules such as tamoxifen and raloxifene may act as strong antagonists in certain tissues (*e.g.*, mammary gland) and agonists in others (*e.g.*, bone).[4] This phenomenon is explained by the fact the specific three-dimensional conformation induced by a given ligand determines the accessibility of co-activators and co-repressors which control the transcriptional activity of the receptor. This 'tripartite' mechanism of action provides ERs with a significant variety of effects on transcription depending on the cell and promoter taken into consideration.[5]

Hence in the screening and study of compounds active through the ERs, it is most relevant to evaluate the effects of the given compound in all tissues. Well

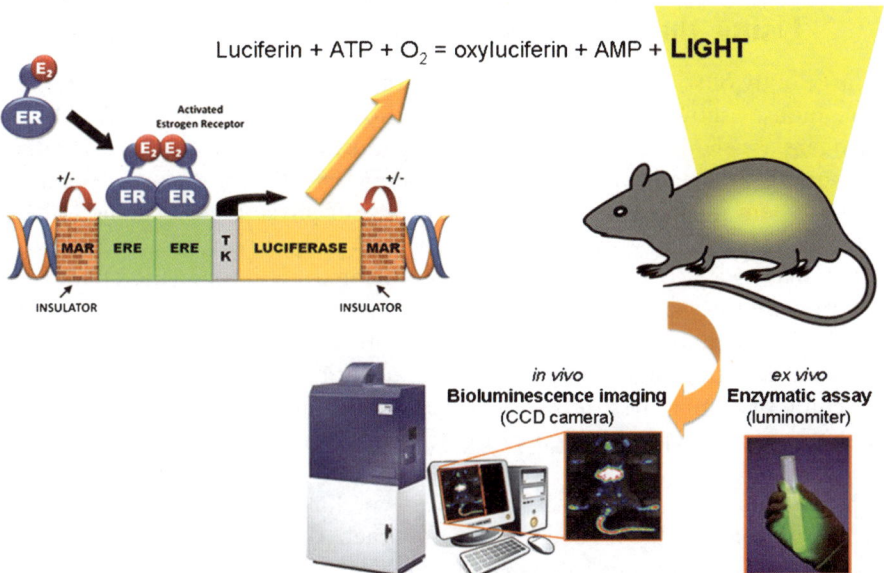

Luciferin + ATP + O$_2$ = oxyluciferin + AMP + **LIGHT**

Figure 8.1 The transgene ERE-Luc integrated in the genome of the ERE-Luc mouse has minimal background activity in the absence of signals activating the ER. Upon ER activation, however, luciferase is produced at high concentration and rapidly accumulates in each ER target tissue. The administration of the exogenous luciferin, substrate of the luciferase, triggers the production of photons which can be measured by a charge-coupled device (CCD) camera. Image analysis software permits the generation of images indicating in pseudocolor the body areas where the photon emission occurs and the intensity of the signal generated.

aware of this necessity, a few years ago we developed a transgenic animal engineered to ensure that an ER reporter gene could be recognized and activated by estrogens in all tissues. To obtain such a model, we randomly integrated in the mouse genome a construct containing a duplicate of the canonical ERE driving the gene encoding the firefly luciferase flanked by insulator sequences (Figure 8.1). The insulator sequences were demonstrated to shield the construct from the influences of the surrounding chromatin and ensured that the transgene is accessible to the activated receptor in all cells.[6,7] As reporter, we selected the natural firefly luciferase gene because of its fast turnover (2–3 h), which is important for the evaluation of the dynamics of ER transcriptional activity and because it is suitable for *in vivo* imaging (Figure 8.1).

The model system we generated, and named ERE-Luc, represents a very useful tool for the spatio-temporal study of ER activity and for the screening and identification of compounds active on the ER, including endocrine disrupters which may be present in the environment or in the alimentary chain.

8.2   Screening of Compounds Active Through the ER Using the ERE-Luc Mouse

A large body of studies carried out by our and others laboratories on the ERE-Luc mouse showed that minimal, physiological changes of circulating estrogens (such as the changes occurring during the estrous cycle) induce measurable and significant changes of luciferase synthesis, thus the sensitivity of the system is such as to highlight the effects of physiological changes of ER activity. The model system is also selective for ER activity because receptors other than ERα or ERβ [*e.g.*, estrogen-related receptors (ERR)] do not functionally interact with the transgene.[6]

After a preliminary set of experiments aimed at finding the concentration and time ensuring the distribution of the luciferin substrate to all organs, the screening of compounds of interest is generally done by an initial time-course analysis of the effect of a high concentration of the compound by measuring bioluminescence at different times (typically 0, 3, 6, 12, 16 and 24 h) in each individual body area (Figure 8.2, left).[8] This preliminary experiment allows for the definition of the time necessary to reach the maximum effect of the compound and allows us to proceed to the analysis of the effect of dosage on the different tissues. For the dose–response relationship, it is advisable to combine the *in vivo* analysis with two sets of *ex vivo* studies: the first is a simple measurement of the extent of bioluminescence from the dissected organs; the second is a biochemical assay aimed at the quantitative measurement of luciferase activity per milligram of protein in each dissected tissue. Generally, the three sets of experimental data are in agreement; however, it is important to bear in mind that the *in vivo* analysis is severely limited by the moderate penetration of the photons through the different layers of tissues and by the

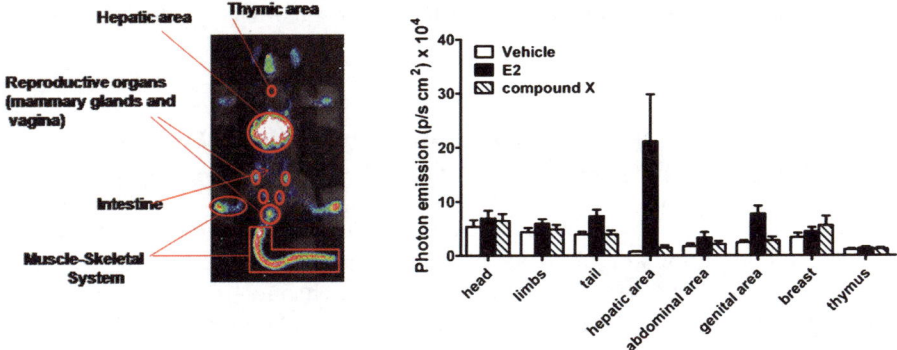

Figure 8.2 *In vivo m*easurement of ER activity in the ERE-Luc mouse. Left: after acquisition of the total photon emission and reconstruction in pseudocolors of the regions where such emission occurs, the analysis can be improved by the definition of more selected body areas in which photon emission is measured. This selection can be done either manually or with the aid of specific algorithms assuring an unbias analysis of bioluminescence. Right: exemplary comparative analysis of the effect of a novel compound (compound X) on ER activity as measured *in vivo* in selected mouse regions. Generally, the activity of novel compounds is measured in relation to the effects of 17β-estradiol (E2): this permits a rapid evaluation of the efficacy of the effect of the compound under study in each tissue.

fact that the two-dimensional analysis is therefore the resultant of the emission of several organs.

8.3 Study of the Effects of Estrogenic Compounds on Brain ER Activity

More challenging may be the evaluation of the effect of the treatments of interest on ER activity in the central nervous system. However, the knowledge of the ability of these compounds to cross the blood–brain barrier is of undoubted interest. The obstacles to these measurements are the inability of the photon produced within the neural tissue to cross the bone of the skull and the limited resolution of the system, which would not allow the definition of the brain areas in which the activity of ER is elicited. To overcome these shortcomings, we attempted to measure luciferase activity *ex vivo* in brain slices.[9] In this case, to ensure its proper distribution, luciferin is administered in the third ventricle at least 20 min prior to tissue dissection and bioluminescence measurement. This methodology proved of interest because it allows the study of ER activity in discrete brain nuclei. This technique, for the first time, made it possible to demonstrate that the estrous cycle significantly influences ER activity in brain areas not related to reproduction, such as limbic areas, and that basal ER activity is high in the brains of both sexes. Further, by showing significant changes in brain ER activity in mice at different stages of the estrous cycle, the studies demonstrated the great sensitivity of the established methodology.

8.4 Spatio-temporal Assessment of the Estrogenic Effects of Natural and Synthetic Compounds

The ERE-Luc mouse model is most useful for the study of long-term effects of estrogenic compounds or mixtures of compounds. In this case, the effects of estrogenic compounds may be monitored daily in chronic treatments or during prolonged exposures (*e.g.*, to diets or to environmental pollutants). The spatio-temporal analysis of the effect of estrogenic compounds can reveal a tissue specificity in the long-term response to treatments. For instance, Figure 8.3 shows that a prolonged treatment with estradiol induces down-regulation of ER response in the hepatic area, but not in genital area, thus indicating that each tissue may respond differently to the same treatment, with significant consequences for potential undesired effects in long-term exposures.

The application of reporter mice, by providing biomedical studies with time as a novel dimension, has the potential to induce a paradigm shift in future toxicological and pharmacological studies. Indeed, the possibility of studying the effect of a given xenobiotic in time in a single individual allows us to evaluate exactly the response to a prolonged exposure to low concentrations of the compound of interest, and the influence on the effects of the exposure to changes in diet, exercise and physiological status (*e.g.*, pregnancy, aging).

The first studies in this sense have already underlined the power of investigations using reporter mice. For instance, we recently investigated the effect of long- term (21 day) hormone replacement therapy on ER signaling in the whole body.[10,11] ER-selective modulators were administered daily at doses equivalent to those used in humans as calculated by an allometric approach. The study demonstrated for the first time that the treatments induced an unexpected oscillation of ER transcriptional activity; interestingly, some of the treatments, in which it is conceivable that the concentration of these compounds was constant due to the daily treatment, induced changes in ER

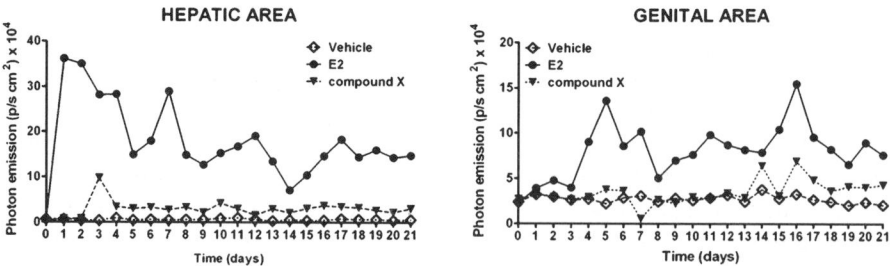

Figure 8.3   *In vivo* analysis of ER activity in the hepatic and genital areas during a 21 day treatment of ERE-Luc mice with 17β-estradiol (E2, 10 μg kg^{-1} per day) or compound X.

activity which reproduced the oscillation observed in cycling animals, where the concentration of estrogens varies in the 4–5 days of the estrous cycle. ER oscillations showed frequencies and amplitudes that were strictly associated with the tissue analyzed and the chemical nature of the compound administered. The application of appropriate algorithms (*e.g.*, agglomerative hierarchical clustering) demonstrated that the spatio-temporal activity of selective ER modulators was predictive of their structure, demonstrating that the analysis of the effect of estrogenic molecules on a single surrogate marker of ER transcriptional activity was sufficient to classify structurally related families of compounds.[10]

Several laboratories applied the ERE-Luc model to the study of the estrogenic effects of natural and synthetic compounds present in the diet and demonstrated that this model provides a novel opportunity to investigate the effect of environmental and alimentary contaminants. For instance, it was demonstrated that in a 21 day treatment, soymilk was more potent than genistein in the liver and appeared to extend its influence on ER transcriptional activity in other tissues, such as the digestive tract. A mixture of pure genistein and daidzein at the same concentration as in soymilk failed to induce significant changes during acute and chronic studies, suggesting an important, uncharacterized role of the soymilk matrix.[12–14] Consistent with this observation, synergistic effects of the matrix plus isoflavones were finally observed in MCF-7 cells stably transfected with the ERE-Luc construct. On the other hand, a study carried out on ERE-Luc mice fed with a bread prepared with Cd-contaminated flour or treated with pure $CdCl_2$ for 21 days confirmed the estrogenicity of $CdCl_2$ but did not suggest the same activity for diet-bound Cd. Both studies suggested caution when extrapolating results from pure compound studies (*e.g.*, estrogenicity of $CdCl_2$) to dietary exposure scenarios.[15]

8.5 Conclusion

The major advantage of reporter animals is that they provide measurable endpoints for the evaluation of drug efficacy or toxic effects in all tissues of living animals. This radically changes animal-based biomedical experimentation and provides the opportunity to understand physiological and pathological mechanisms better and to evaluate and compare the effects of therapeutic treatments in healthy animals or in models of specific diseases. However, most of the reporter animals so far generated enable us to measure the effects of the compounds of interest on a single molecular event. Moreover, the availability of very sophisticated reporters for bioluminescence (genetic modification of luciferase and luciferases of diverse biological origin), fluorescence (genetic modifications of GFPs creating molecules with emission at a variety of wavelengths) and radiochemical analysis will facilitate the generation of animal models in which more than one molecular process can be studied at the same time (*e.g.*, analysis of proliferation and apoptosis at the same time, proliferation and inflammation and so on).[16–18] These models would represent an invaluable model system for the rapid definition of the specific effects of a given compound on its specific target and, in addition, provide a more general view on the overall pathophysiological consequences of exposure to a specific compound.

Finally, the main species used in the generation of reporter animals so far has been the mouse. Reporter-gene technology, however, should be extended to species other than mice and might also eventually include humans. The generation of viral vectors able to carry the reporter gene and infect species other than rodents would represent a major step forward: being able to visualize in real time the effects of any given compound would mean that we could tailor any treatment to specific needs with a tremendous increase in the efficiency of drug discovery and toxicological research.

Acknowledgements

The work in this laboratory is supported by grants from the European Community (INMiND) and by Pfizer.

References

1. K. Dahlman-Wright, V. Cavailles, S. A. Fuqua, V. C. Jordan, J. A. Katzenellenbogen, K. S. Korach, A. Maggi, M. Muramatsu, M. G. Parker and J. A. Gustafsson, *Pharmacol. Rev.*, 2006, **58**, 773.
2. J. W. Thornton, E. Need and D. Crews, *Science*, 2003, **301**, 1714.
3. G. V. Markov and V. Laudet, *Mol. Cell. Endocrinol.*, 2011, **334**, 21.
4. P. Huang, V. Chandra and F. Rastinejad, *Annu. Rev. Physiol.*, 2010, **72**, 247.
5. J. A. Katzenellenbogen, B. W. O'Malley and B. S. Katzenellenbogen, *Mol. Endocrinol.*, 1996, **10**, 119.
6. P. Ciana, M. Raviscioni, P. Mussi, E. Vegeto, I. Que, M. G. Parker, C. Lowik and A. Maggi, *Nat. Med.*, 2003, **9**, 82.
7. A. Maggi and P. Ciana, Italian Patent ITA MI001503; European Patent EP 1298988B1; US Patent 7 943 815, 2002.
8. P. Ciana, A. Biserni, L. Tatangelo, C. Tiveron, A. F. Sciarroni, L. Ottobrini and A. Maggi, *Mol. Endocrinol.*, 2007, **21**, 388.
9. A. Stell, S. Belcredito, P. Ciana and A. Maggi, *Mol. Imaging*, 2008, **7**, 283.
10. G. Rando, D. Horner, A. Biserni, B. Ramachandran, D. Caruso, P. Ciana, B. Komm and A. Maggi, *Mol. Endocrinol.*, 2010, **24**, 735.
11. S. Della Torre, A. Biserni, G. Rando, G. Monteleone, P. Ciana, B. Komm and A. Maggi, *Endocrinology*, 2011, **152**, 2256.
12. G. Rando, B. Ramachandran, M. Rebecchi, P. Ciana and A. Maggi, *Toxicol. Appl. Pharmacol.*, 2009, **237**, 288.
13. M. Penza, C. Montani, A. Romani, P. Vignolini, B. Pampaloni, A. Tanini, M. L. Brandi, P. Alonso-Magdalena, A. Nadal, L. Ottobrini, O. Parolini,

E. Bignotti, S. Calza, A. Maggi, P. G. Grigolato and D. Di Lorenzo, *Endocrinology*, 2006, **147**, 5740.
14. C. Montani, M. Penza, M. Jeremic, G. Biasiotto, G. La Sala, M. De Felici, P. Ciana, A. Maggi and D. Di Lorenzo, *Toxicol. Sci.*, 2008, **103**, 57.
15. B. Ramachandran, S. Mäkelä, J. P. Cravedi, M. Berglund, H. Håkansson, P. Damdimopoulou and A. Maggi, *Toxicol. Lett.*, 2011, **202**, 75.
16. N. Thorne, J. Inglese and D. S. Auld, *Chem. Biol.*, 2010, **17**, 646.
17. A. Miyawaki, *Annu. Rev. Biochem.*, 2011, **80**, 357.
18. S. Mehta and J. Zhang, *Annu. Rev. Biochem.*, 2011, **80**, 375.

CHAPTER 9

From Computational Simulations on Nuclear Receptors to Chemosensors for Food Safety

CHIARA DALL'ASTA*[a], ANDREA FACCINI[b] AND GIANNI GALAVERNA[a]

[a] Department of Food Science, University of Parma, Viale Usberti 17/A, 43124 Parma, Italy; [b] Services and Research Center, CIM Laboratory 'Centro Interdipartimentale Misure – Giuseppe Casnati', University of Parma, Parma Technopole, Parco Area delle Scienze, 43100 Parma, Italy
*E-mail: chiara.dallasta@unipr.it

9.1 Cyclodextrins: an Overview

Molecular recognition is a very fundamental process in our world since enzymes, antibodies, membranes and their receptors, carriers and channels all work through this principle.[1,2]

As a pioneer in this field, Fischer in 1984 proposed the *lock and key* model,[3] according to which the two actors of the molecular recognition process, the molecular receptor and the substrate to be recognized, may be represented as the lock and the key to give a defined receptor–substrate complex. Chemists have fully exploited such a model, designing synthetic systems with fascinating properties observed in natural systems and also creating novel organic chemistry of great interest to both science and technology. It was in the mid-1980s that molecular recognition became a new and productive area of chemistry, certainly based on the impressive achievements of organic synthesis

RSC Drug Discovery Series No. 30
Computational Approaches to Nuclear Receptors
Edited by Pietro Cozzini and Glen E. Kellogg
© The Royal Society of Chemistry 2012
Published by the Royal Society of Chemistry, www.rsc.org

such as those pertaining the total synthesis of complex natural products, but developing a new philosophy for science and thinking called *supramolecular chemistry*, the chemistry beyond the molecule,[4-6] where non-covalent bonds and spatial fit between molecular individuals that form a specific host–guest complex are in the foreground.[7,8] According to this general approach, molecular recognition is defined as the study of multi-molecular entities and assemblies called supramolecular complexes, formed between two or more chemical species held together by non-covalent forces.[9]

Among the very huge number of potential hosts, from relatively simple receptors containing one or more binding sites to dipodal, tripodal receptors, clips, appropriate functionalized cyclophanes, crown ethers, azamacrocycles, *etc.*, cyclodextrins (CDs) are one of the most important classes of compounds, for the following reasons:[10]

- They are natural products, produced from starch by a relatively simple enzymatic conversion.
- They are produced in thousands of tons per year by means of environmentally friendly technologies.
- Owing their ability to give inclusion complexes, important properties of the complexed compounds can be modified significantly.
- Any toxic effect is secondary and can be eliminated by selecting the appropriate CD type or derivative or mode of application.
- As a result of the previous point, CDs are largely consumed by humans as ingredients in drugs, foods and/or cosmetics.

9.1.1   Structure and Properties of CDs

CDs are a family of cyclic oligomers composed of α-(1→4)-linked D-glucopyranose units in a $^4C_1$ chair conformation. This peculiar structure gives rise to the most notable characteristic: the presence of a hydrophobic conical cavity in the frame of an essentially hydrophilic compound, which is at the basis of their widespread applications in different fields of chemistry.

The most common CDs have six, seven and eight glucopyranose units and are referred to as α-, β- and γ-cyclodextrin, respectively. Larger cyclodextrins have also been identified, isolated and also synthesized for different applications.[11,12] The cavity is limited by hydroxyl groups of different chemical character. Those located at the narrower side belong to the C6 of the glucopyranose ring (primary side, upper rim) and are primary hydroxyls, whereas those located at the wider entrance are secondary hydroxyls and therefore are less prone to chemical transformation (secondary side, lower rim). The reactivity of the hydroxyl groups strongly depends on the reaction conditions, cyclodextrins generally behaving as polyols.

Figure 9.1 shows the characteristic structural features of cyclodextrins. The cavity is wider on the lower rim where the secondary hydroxyl groups are located, whereas on the upper rim the free rotation of the primary hydroxyls reduces the effective diameter of the cavity.

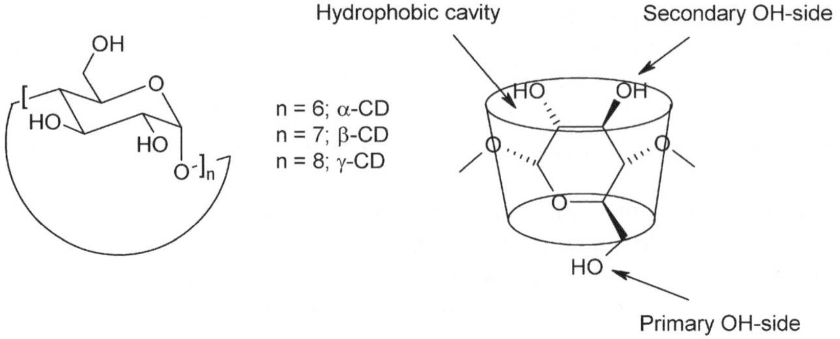

Figure 9.1   Structure of cyclodextrins.

The inner diameter of the cavity in unmodified cyclodextrins varies from 5 to 10 Å and it is about 8 Å in depth. The non-bonding electron pairs of the glucosidic oxygen bridges are directed towards the inside of the cavity, generating a high electron density and giving some Lewis base character to the environment.

Among the three most important compounds, β-cyclodextrin is characterized by a fairly rigid structure, on account of a complete secondary belt of hydrogen bonds formed between the C-2 OH group of one glucopyranoside unit and the C-3 OH group of the adjacent glucopyranose unit. This intramolecular H-bond formation is probably responsible for the lower solubility of β-cyclodextrin in water. On the other hand, the H-bond belt is incomplete in the α-cyclodextrin molecule, because one glucopyranose unit is in a distorted position. Consequently, instead of six possible H-bonds, only four can be established simultaneously. The γ-cyclodextrin, forming less intramolecular hydrogen bonding, is characterized by a more flexible structure, thus being also the most soluble CD.

The most important characteristics of cyclodextrins are summarized in Table 9.1.

Table 9.1   Properties of cyclodextrins.

Property	α-CD	β-CD	γ-CD
No. of glucose units	6	7	8
MW	972	1135	1297
Solubility in water (g per 100 mL at room temperature)	14.5	1.85	23.2
$[\alpha]_D$ (25 °C)	150 ± 0.5	162.5 ± 0.5	177.4 ± 0.5
Cavity diameter (Å)	4.7–5.3	6.0–6.5	7.5–8.3
Height of torus (Å)	7.9 ± 0.1	7.9 ± 0.1	7.9 ± 0.1
Diameter of outer periphery (Å)	14.6 ± 0.4	15.4 ± 0.4	17.5 ± 0.4
Approx. volume of cavity (Å^3)	164	262	427
Crystal water (wt%)	10.2	13.0–14.5	8.13–17.7

In the early 1950s and during the past decades, a series of the larger CDs were isolated and studied.[13] For example, the nine-membered δ-CD was isolated from the commercially available CD conversion mixture by chromatography. δ-CD showed a greater aqueous solubility than β-CD, but less than α- and γ-CD. It was the least stable among the CDs known at that time: the hydrolysis rate increases in the order α-CD < β-CD < γ-CD < δ-CD. Moreover, δ-CD did not show any significant complexation effect on slightly soluble drugs in water, except in the case of some large guest molecules, probably on account of a collapsed structure not showing a regularly cylinder-shaped cavity and thus of smaller dimensions, in agreement also with computer graphics studies. Owing to the high solubility, low viscosity and the inability to undergo retrogradation, a number of industrial applications have been suggested for these novel starches: in the food industry as a high-energy additive to soft drinks, as a retrogradation retardant in breads, for bread improvement, freeze-resistant jellies and for the production of non-sticky rice. Furthermore, several non-food applications have been suggested, including paper coating materials and as starch substitutes in adhesives and biodegradable plastics.

9.1.2 CD Inclusion Complexes

CDs, having the form of an empty toroidal capsule, are able to form so-called *inclusion complexes*, including another molecule in their internal cavity.[14] Figure 9.2 e illustrates the formation of a CD inclusion complex.[15]

As a general definition, inclusion complexes are entities involving two or more molecules in which one of the molecules, the 'host', includes, totally or in part, a 'guest' molecule, solely by physical interactions, that is, without covalent binding. CDs are typical host molecules and may include a great variety of molecules with dimensions fitting their cavity; typically, inclusion complexes are formed with molecules having the size of one or two condensed benzene rings, or even larger ones which have a side chain of comparable size.

Complex formation is governed by the dimensional fit between the cavity and the guest molecule: the lipophilic cavity of CDs can include appropriately sized non-polar moieties.[16] Formation of inclusion complexes does not involve the formation or breaking of covalent bonds.[17] The release of enthalpy-rich water molecules from the cavity is the main driving force for complex

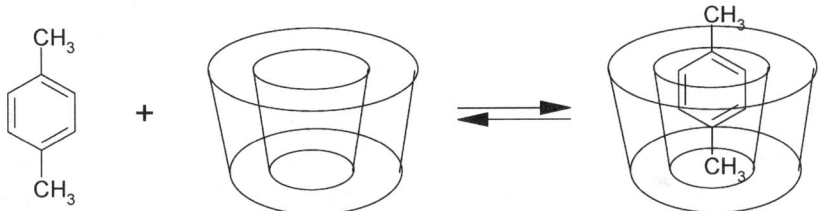

Figure 9.2 Schematic illustration of inclusion complexation of *p*-xylene by a CD.

formation; water molecules are displaced by more hydrophobic guest molecules, thus allowing for hydrophobic interactions between the guest and the CD cavity.

The binding of guest molecules within the host CD may be described as a dynamic equilibrium in which the binding strength depends on the host–guest best fitting coupled to specific interactions between surface atoms.

The physicochemical properties of guest molecules are greatly influenced by the inclusion as they are temporarily locked or caged within the CD cavity, giving rise to large modifications of guest molecule properties.[18]

Among the most important properties, the following should be considered:

- solubility enhancement of highly insoluble guests;
- stabilization of labile guests against the degradative effects of oxidation, visible or UV light and heat;
- control of volatilization and sublimation;
- physical isolation of incompatible compounds;
- modification of the chromatographic behavior.

The potential guest list for inclusion complex formation is fairly varied and includes several different classes of compounds, such as straight- and branched-chain aliphatics, aldehydes, ketones, alcohols, organic acids, fatty acids, aromatics, gases and polar compounds such as halogens, oxy acids and amines.[19]

Inclusion complex formation is based on two key factors. The first is the steric balance between the size of the CD and the size of the guest molecule, which has to fit properly into the CD cavity. The second is the thermodynamic interactions between the different components of the system (CD, guest, solvent): there must be a favorable net energetic driving force to favor the inclusion of the guest inside the cavity.

Four energetically favorable interactions are involved in the process of inclusion complex formation:

- the displacement of polar water molecules from the apolar CD cavity;
- the increased number of hydrogen bonds formed as the displaced water returns to the larger pool;
- a reduction in the repulsive interactions between the hydrophobic guest and the aqueous environment;
- an increase in the hydrophobic interactions as the guest inserts itself into the apolar CD cavity.

Conformational adjustments of the guest may also take place inside the cavity in order to fit the steric requirements perfectly. Inclusion complexes are mostly formed with a host:guest ratio of 1:1, although 2:1, 1:2, 2:2, or even more complicated associations and higher order equilibria also exist.

Complex formation can be described using the classical equilibrium equation governed by the complex stability constant K_a, in agreement with the thermodynamic equilibrium between the host (CD) and the guest molecule (D) (Figure 9.3).

$$CD + G \rightleftharpoons CD*G$$

$$K_a = \frac{[CD*G]}{[CD][G]}$$

Figure 9.3 Equilibrium of complexation [eqn (9.1)] and the complex stability constant K_a [eqn (9.2).

The formation of the inclusion complexes in aqueous solution has several consequences:

- The concentration of the guest in the dissolved phase increases greatly owing to increase in solubility.
- The spectral properties of the guest are modified. For example, the chemical shifts of the anisotropically shielded atoms are modified in the NMR spectra. Also, when achiral guests are included in the chiral cavity, they become optically active and show strong induced Cotton effects on the circular dichroism spectra. Further, the maxima of the UV spectra are shifted by several nanometers, and also the fluorescence is very strongly improved, since the fluorophore guest is protected from the quenching effect exerted by waters or other quenchers.
- The reactivity of the included molecule is modified. In most cases, the reactivity decreases since the guest is protected, but in many cases CDs may behave as artificial enzymes, catalyzing several chemical reactions and modifying the reaction pathways.
- The polarity of the complex is completely different from that of the guest, being much more hydrophilic and, thus, showing a completely changed chromatographic mobility, hence with potential exploitation for analytical purposes.

As far as the determination of the equilibrium constant is concerned, a wide variety of spectroscopic methods are frequently employed, on the basis of the spectral properties of the guest. For aromatic guests, the strong absorption bands in the UV–Vis region are used, whereas in the case of aliphatic guests which do not show any intense absorption bands in the UV–Vis region, chromophoric compounds (*e.g.*, azo dyes) are added to the CD solutions in a competitive test format and the binding constants are determined by differential UV spectroscopy.[20]

In the case of fluorescent guests, the association constant can be determined by exploiting emission change of the substrate with the CD concentration according to the Benesi–Hildebrand model.[21]

NMR titration is also used to obtain the thermodynamic quantities for CD complexation. A mathematical approach similar to that employed in absorption and fluorescence spectroscopic determinations is used. NMR spectroscopy is widely used in chiral discrimination studies using CDs.[22–25]

Circular dichroism spectroscopy is one of the best methods for the observation of the complexation behavior of chromophoric guests with CDs, since all natural CDs are inherently chiral and the spectral changes caused by the inclusion of guest molecules are often more enhanced in the circular dichroism spectra than in the UV spectra.[26–28]

More recently, the use of scanning probe techniques such as atomic force microscopy has allowed measurements of the force involved in the inclusion processes at a single-molecule level, opening up new and exciting prospects in supramolecular chemistry.[29–31]

9.2 Cyclodextrins as Chemosensors: Recognition Based on Spectral Changes

9.2.1 Modification of Absorption Spectra by CDs

The high electron density prevailing inside the CD cavity affects the electrons of the incorporated molecules. This results in characteristic changes in various spectral properties of both the host and the guest, which have been exploited for the design of chemosensors for several analytical applications.[32] Since the spectral changes of colored molecules in the presence of CDs was first observed in 1951 by Cramer,[33] the effect of CDs on the UV and visible spectra of various guest molecules has been studied.[34,35]

Some of the most useful effects are as follows:

- increased solubility of apolar analytes and/or reagents in aqueous media;
- increased stability of sensitive reagents and the color complexes in aqueous or non-aqueous solutions;
- increased sensitivity of the colored reaction through intensification of UV absorption;
- improved selectivity of colored reactions.

These effects make CDs useful auxiliaries in the spectrophotometric determinations of a wide variety of compounds and elements.[36]

Generally, a bathochromic shift and an absorbance change (increase or decrease) of the guest molecule can be observed upon complexation with CDs. The changes in absorbance have been used to calculate the association constants.

9.2.2 Induced Circular Dichroism

By adding a CD to an aqueous solution of a potential achiral guest, an induced Cotton effect will be observed on the circular dichroism spectra, which can be attributed to the optical activity of the guest molecule induced by inclusion in the CD chiral cavity and partly to conformational changes of the same cavity.

The Cotton effect is only observed when the guest molecule, or more exactly its chromophore moiety, is really included in the CD cavity. An outer surface association of a potential guest with the CD molecule may lead to some modification of the spectral properties, but not to induce circular dichroism.[37]

It has been found that the sign and the intensity of the induced Cotton effects are sensitive to the orientation of the guest chromophore in the CD cavity. In particular, Cotton effects with opposite signs will be observed depending on whether the electric dipole moment of the guest coincides with the axis of the CDs or, instead, if they are perpendicular to each other.[38]

9.2.3 Fluorescence Enhancement

Molecular fluorescence spectrophotometry is a routine technique in many analytical applications, in particular for its lower detection limit and greater selectivity compared with molecular absorption spectrophotometry. However, although most compounds show strong fluorescence in organic solvents, the intensity is rather weak in water, which acts as a quencher. Adding CDs, which form inclusion complexes with analyte molecules in aqueous solutions, can result in significant fluorescence enhancement.[39]

The first utilization of CDs in fluorescence enhancement was by Kinoshita *et al.*, who examined the effect of CDs on dansylamino compounds.[40]

The inclusion of analyte molecules in the CD cavity produces several effects:

- The CD cavity protects the fluorescing exciting state of the analyte from external quenchers.[41,42]
- As a consequence of inclusion complex formation, the rotation of the guest molecule is hindered and the relaxation of the solvent molecules is considerably decreased. Both of these effects can result in a decrease in the fluorescence vibrational deactivation.
- The CD cavity behaves similarly to the organic solvent, affording an apolar surrounding for the included chromophore. This altered environment can provide favorable polarity for enhanced quantum efficiencies and hence the intensities of luminescence. The effective microenvironment of the CD cavity is likely to be similar to that of oxygenated solvents such as dioxane, *tert*-amyl alcohol and 1-octanol.[43,44]
- The CD solution can improve the detection limit for hydrophobic analytes in aqueous solution by increasing their solubility.

Accordingly, it has been found that the fluorescence intensities of many compounds, such as pyrene,[45] drugs, narcotics, hallucinogenics,[46] and polychlorinated biphenols,[47] are significantly increased by complex formation with CDs and their derivatives.

Another important application of CDs in the field of analytical luminescence chemistry is the development of fiber-optic CD-based (FCD) sensors for the detection of a widely variety of organic compounds.[48]

9.3 Cyclodextrin Derivatives: Strategies for Chemical Modification

β-CD is the most widely used and represents at least 95% of all CDs produced and utilized, for several reasons (price, availability, approval status, cavity dimensions, *etc.*), although its anomalous low aqueous solubility places some limits on its wider utilization. Fortunately, the solubility of all CDs can be improved markedly by chemical or enzymatic modifications.

Considering that CDs contain 18 (α-CD), 21 (β-CD) and 24 (γ- CD) hydroxyl groups, the number of possible derivatives is virtually unlimited, and so far more than 1500 derivatives have been synthesized. Hydroxyl groups can be modified by replacing the hydrogen atom or the hydroxyl group with a variety of different functions, such as alkyl, hydroxyalkyl, carboxyalkyl, amino, thio, tosyl, glucosyl, maltosyl, and thousands of ethers, esters, anhydro, deoxy, acidic, basic, *etc.*, derivatives can be prepared by chemical or enzymatic reactions. The aim of such derivatizations may be to improve the solubility of the CD derivatives (and their complexes), to improve the fitting and/or the association between the CD and its guest, to attach specific (catalytic) groups to the binding site (*e.g.*, in enzyme modeling) or to form insoluble, immobilized CD-containing structures and polymers (*e.g.*, for chromatographic purposes).

The actual or potential uses of native CDs or their derivatives in pharmaceuticals, foods, cosmetics, chemical products and technologies have been widely summarized in many reviews,[49–52] and also in some CD monographs, and illustrate the steady increase of the CD market during the last decades.

Through modifications, CDs can provide new molecules that can have different abilities ranging from enzyme-like activity to receptor-like binding.[53,54] The strategy for modifications depends on the purpose of the final product and many factors, such as the number of substituents, the regiochemistry of the substitution and the stereochemical changes that have taken place during the synthesis, have to be established in order to address the synthetic strategy.

9.3.1 The Chemistry for Modification of CDs

Two primary factors need to be considered in the chemistry of CDs for their chemical derivatization or modification: the nucleophilicity of the hydroxyl groups and the ability of CDs to form complexes with the reagents used.

Since hydroxyl groups are nucleophilic, the initial reaction, which controls the regioselectivity and the extent of modification (mono, di, tri, *etc.*) of all subsequent reactions, is an electrophilic attack on these positions.

Of the three types of hydroxyl groups present in CDs, those at the 6-position are the most basic (and often most nucleophilic), those at the 2-position are the most acidic and those at the 3-position are the most inaccessible.[55–57]

As a general rule, the more reactive the reagents are, the less selectively the hydroxyl group will be funtionalized. Thus, highly reactive reagents will react not only with hydroxyl groups at the 6-positions but also with those on the secondary side, whereas less reactive reagents will react more selectively with the hydroxyl groups at 6-positions.

An interesting factor affecting the chemistry of the hydroxyl groups is provided by the ability of CDs to form complexes. If the complex formed is very strong then the predominant product formed will be dictated by the orientation of the reagent within the complex. On the other hand, if the complex is weak, then the product formation will be directed by the relative nucleophilicities of the hydroxyl groups.

It is worth noting that both the solvent and the size of the CD cavity play important roles in determining the strength and the orientation of the complex between the reagent and the CD, and also in affecting the product of the reaction.

Thus, a general strategy is to protect some hydroxyl groups selectively and direct the incoming reagent exclusively to the other free hydroxyl groups.

Although polyfunctionalized CD derivatives are very important in organic chemistry, their synthesis is fairly difficult, on account of the inevitable formation of a mixture of different products that have to be separated by tedious chromatographic techniques such as RP-HPLC.

9.4   Cyclodextrins as Chemosensors for Food Diagnostics

In the field of analytical chemistry, CDs and their derivatives have been extensively used in separation methods (chromatography, electrophoresis) for their ability to improve separations and to discriminate between positional isomers, functional groups, homologs and enantiomers.[58,59] They have been used both as additives to the mobile phases and to prepare bonded stationary phases which have shown a large number of applications and have been applied to the preparation of commercially available chromatographic columns.

Moreover, CDs have been applied in several methods based on fluorescence spectroscopy for their ability to increase the fluorescence intensity of a guest molecule upon inclusion in the hydrophobic cavity by protecting it from the quenching effect exerted by water.

On account of these properties, CDs have been exploited for the development of a number of analytical methods with applications in the fields of nanotechnology, pharmacology, material science and food chemistry. In particular, CDs can be efficiently exploited as chemosensors for developing diagnostic tools for food safety. An example of this application is given in the following section, regarding the use of CDs for mycotoxin determination in food.

9.5   Cyclodextrin–Mycotoxin Complexation: a Case Study in the Food Safety Field

9.5.1   Mycotoxins: General Information

The term 'mycotoxin' is usually reserved for the toxic chemical products formed by a few fungal species that readily colonize crops in the field or after harvest and thus pose a potential threat to human and animal health through the ingestion of food products prepared from these commodities.

Each mycotoxin is produced by one or more very specific fungal species. In some cases, one species can produce more than one mycotoxin. For example, the aflatoxins can be produced by *Aspergillus flavus*, *Aspergillus parasiticus* and limited other *Aspergillus* species, whereas ochratoxin A is considered to be mainly the product of *Aspergillus ochraceus* in tropical regions and *Penicillium verrucosum* in temperate areas. However, the presence of a recognized toxin-producing fungus does not automatically imply the presence of the associated toxin as many factors are involved in its formation. Conversely, the absence of any visible mould does not guarantee the absence of toxins as the mould may have already died out while leaving the toxin intact.

Mycotoxins can occur both in tropical areas and in temperate regions of the world, depending on the species of fungi. Major food commodities affected are cereals, nuts, dried fruit, coffee, cocoa, spices, oil seeds, dried peas and beans and fruit, particularly apples. Mycotoxins may also be found in beer and wine resulting from the use of contaminated barley, other cereals and grapes in their production. Mycotoxins also enter the human food chain *via* meat or other animal products such as eggs, milk and cheese as the result of livestock eating contaminated feed.

Mycotoxins cause a diverse range of toxic effects because their chemical structures are very different from each other. Acute effects require that high amounts are present when eaten so that such incidents are usually restricted to people in the less developed parts of the world where resources for control are limited, or to livestock. Chronic effects are of concern for the long-term health of the human population and are important when the mycotoxins are present in much lower amounts. Some of the most common mycotoxins are carcinogenic or genotoxic or may target the kidney, liver or immune system.

National and international organizations are constantly evaluating the risk that such mycotoxins pose to humans. For some mycotoxins this has resulted in statutory or guideline maximum permissible limits. Many countries now have legal limits for aflatoxins, which are the most widespread and toxic mycotoxins, although these values are by no means uniform.

Most mycotoxins are chemically stable so they tend to survive storage and processing even when cooked to fairly high temperatures such as those reached during baking bread or producing breakfast cereals. This makes it important to avoid the conditions that lead to mycotoxin formation as far as possible. Most mycotoxins are toxic in very low concentrations, so this requires sensitive and reliable methods for their detection. Sampling and analysis taken together represent an extremely demanding challenge for the analyst. Failure to achieve a satisfactory performance can lead to unacceptable consignments being accepted or satisfactory loads being unnecessarily rejected. Cargoes may be extremely valuable and disputes about the analytical results can be extremely costly, especially if this results in the reputation of the trader or retailer being unfairly tarnished. The difficulty of removing a mycotoxin once formed means that the best method of control is prevention.

Among the mycotoxins known to have a native fluorescence emission are certain of the aflatoxins, ochratoxins, zearalenone and related congeners. The structures, toxic effects and fluorescence characteristics of these three groups of toxins vary widely. The fluorophores are derived from a variety of molecular structures. The aflatoxins contain the coumarin moiety, which is highly fluorescent. Similarly, the ochratoxins contain an isocoumarin moiety linked to the amino acid phenylalanine, and zearalenone and related compounds are resorcylic acid lactones (Figure 9.4).

Even small changes in a toxin's structure can dramatically influence the fluorescence. The fluorescence emission of certain of the aflatoxins, those containing a double bond in the furan moiety, can be enhanced by a number of techniques, including halogenation, hydrolysis and rearrangement to the more fluorescent phenolate ions, or photochemical reaction. The fluorescence emission of many fluorophores is also known to be sensitive to the local environment and this is also true for the aflatoxins, ochratoxin A (OTA) and zearalenone (ZEN). The fluorescence emission of aflatoxins B$_1$ and G$_1$ (AFB1 and AFG1) are substantially greater in solvents such as methanol and chloroform than in water.[60] The emission maximum is also influenced by solvation, with a shift towards shorter wavelengths (blue shift) as the environment becomes less polar, an effect observed for AFB1 in various solvents.[61]

The use of CDs for mycotoxin analysis is a relatively recent research field and few studies on mycotoxin–CD interactions have been reported so far. The most extensively studied complexes are those formed by aflatoxins and β-CDs, which have been investigated by means of both spectroscopic and chromatographic techniques.[62–66] More recently, the interaction of zearalenone with β-CD has been completely elucidated using spectroscopic methods (fluorescence

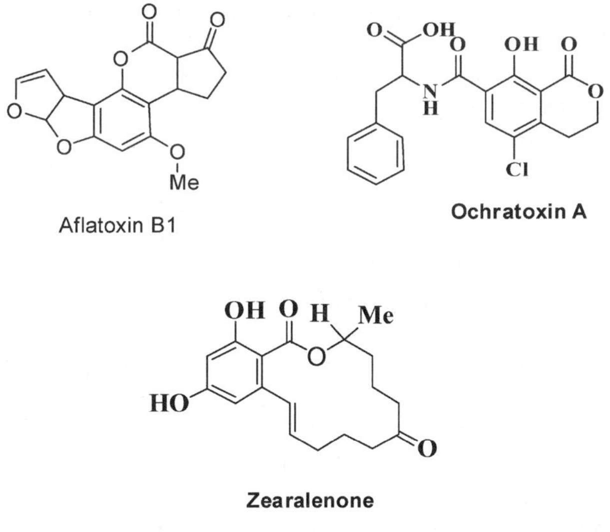

Aflatoxin B1

Ochratoxin A

Zearalenone

Figure 9.4 Structures of aflatoxin B$_1$, ochratoxin A and zearalenone.

spectroscopy, one- and two-dimensional NMR, mass spectrometry).[67] Conversely, the hypothesized interaction between ochratoxin A and β-CD is more controversial and further studies are required for this mycotoxin.[68]

9.5.2   Applications of CDs to the Analysis of Mycotoxins Having Native Fluorescence

The aflatoxins have a native fluorescence that can be excited with UV light (360–365 nm).

The effects of CDs on substituted coumarins, such as the aflatoxins, have been reported in a large number of studies.[61,69] Among these, the most interesting applications include the use of CDs in fungal culture media as an aid in the identification of toxin-producing isolates and the use of CDs as fluorescence enhancers in the determination of aflatoxins, mainly by using HPLC or electrophoretic assays.[62,63,69–73]

Several CDs have been screened for their ability to enhance aflatoxin fluorescence and many of these caused a blue shift in the emission from 440 to 435 nm, suggesting the formation of an inclusion complex through the involvement of the furan moiety.[63] The greatest relative enhancement was observed with succinyl-β-CD, but others such as dimethyl-β-CD (DIMEB) and carboxymethyl-β-CD were also fairly effective.

Based on models of the docking of AFB1 with β- and γ-CDs, it has been suggested that inclusion of the fluorophore in the CD cavity may reduce the quenching effect of the solvent, thereby enhancing fluorescence.[64] Moreover, modeling studies have also suggested that the dihydrofuran portion of AFB1 is inserted into β-CD, with possible hydrogen bonding between the carbonyl groups of the toxin and the secondary hydroxyl groups of the CD.[64]

OTA is a substituted isocoumarin mainly produced by *A. ochraceus* and *P. verrucosum*. The fluorescence emission of OTA is sensitive to the hydrophobicity and pH of the environment.[74,75] The absorption spectrum of OTA changes dramatically with pH, with the band near 320 nm decreasing and the band near 370 nm increasing with increase in pH over the range 3.5–11.[68]

The effect of β-CD on the spectroscopic properties of OTA in aqueous solution was investigated by Verrone *et al.*[68] A 1:1 stoichiometry of OTA/β-CD was observed at all tested pHs (range 3.5–9.5) with an increase in emission intensity of up to about twofold (excitation at 330 nm, emission at 450 nm). It was reported that β-CD interacted with both the protonated and deprotonated forms of OTA, although the dianionic form of OTA seemed to interact more strongly than the protonated form with β-CD. Molecular modeling simulations have also suggested that the phenylalanine portion of OTA is involved in the inclusion complexation with β-CD.[64]

Although the effect of β-CD on OTA fluorescence intensity is not as large as it is for the aflatoxins, the CDs have still been found to be useful in chromatographic assays for increasing the separation from other interferences.[72]

The interaction between ZEN and CDs has already been used for electrophoretic separation of this mycotoxin, also showing a good fluorescence enhancement in the presence of β-CD.[76] The complexation mechanism was recently elucidated by Dall'Asta *et al.*, using for the first time both spectroscopic studies and one- and two-dimensional NMR experiments.[67] The collected data allowed for the definitive elucidation of the complex structure, demonstrating the inclusion of the phenolic moiety and also the double bond in the β-CD cavity. Experiments were also performed for the metabolic derivatives α- and β-zearalenols (ZOLs), showing a similar complexation mechanism, although the reduction of the keto group induced a distortion in the molecular structure, thus lowering the affinity for the CD cavity on account of the higher steric hindrance (C. Dall'Asta *et al.*, unpublished results).

Moreover, a systematic description of the effects of 22 CDs on ZEN fluorescence was reported.[76] The CDs giving the greatest enhancement of fluorescence were DIMEB, 6-monodeoxy-6-monoamino-β-CD, carboxyethylated-β-CD and β-CD.

The possibility of using CDs for the emission enhancement of fluorescently derivatized mycotoxins was reported by Maragos *et al.*[77] This interesting approach is based on the choice of a CD-fitting fluorescent probe for the derivatization of a non-fluorescent mycotoxin. In particular, the derivatization of T-2 toxin with pyrene-1-carbonyl cyanide and also with 1-anthroylnitrile was proposed: the fluorescent labeling allowed for a more sensitive detection of the target compounds. The stoichiometry and the binding constant of the T2–Pyr:CD inclusion complex were determined by means of the Benesi–Hildebrandt equation. In the range of concentrations studied, a 1:2 T2-Pyr:CD stoichiometry was found; this result is consistent with the common behavior of pyrene, which is usually complexed by two CDs.

This sandwich-like complex allows strong fluorescence enhancement, the fluorophore being highly protected from the water quenching effect. The binding constant showed high values ($\log K = 3.88$), suggesting a good interaction between the T-2 derivative and the CD. Also in this case the results are in agreement with the well-known affinity of pyrene for the CD cavity.

9.6 Cyclodextrins as Selectors for the Development of Rapid Fluorimetric Assays

On account of the need for rapid and simple testing assays to be used for mycotoxin detection, several applications have been reported based on the use of CDs as chemosensors for enhancing the native fluorescence of mycotoxins. The main fields of application are the differentiation between toxinogenic and non-toxinogenic fungal strains and the rapid detection of mycotoxins in key food commodities such as milk.

Several studies aimed at the differentiation between aflatoxin-producing and non-producing strains of the *A. flavus* group were based on the addition of

CDs to culture media to allow for more sensitive detection of aflatoxinogenic strains.[78–80] Although very sensitive, these studies suffered from false positives (20%). Rojas-Durán *et al.*[81] demonstrated the first application of room temperature phosphorescence (RTP) to determine aflatoxinogenic strains by use of CDs. Compared with fluorimetry, RTP offers some important advantages: the use of phosphorescent probes allows the removal of the high fluorescent backgrounds observed in *in vivo* measurements. Moreover, phosphorescent indicators offer chemical stability, long triplet lifetimes and RTP excitation spectra in the near-infrared region, where biological tissues adsorb very little. In the study, no significant interferences by non-toxinogenic strains or any of their metabolites were observed and good correlation was achieved between the RTP method and HPLC analysis.

As far as the use of CDs as selectors for the rapid detection of mycotoxins is concerned, the few published studies dealt with the detection of aflatoxins in food without using immunoaffinity clean-up (IAC) and enrichment. Aghamohammadi and Alizadeh[65] proposed the determination of AFB1 in pistachios using synchronous fluorimetry in combination with multivariate calibration methods and derivative techniques. This method was also based on the use of β-CD in a methanol–water solution (10:90, v/v) and gave a good correlation for the quantification of AFB1 without IAC.

Similarly, a modified fluorimetric assay was proposed[82,83] for the determination of AFM1 in milk without IAC. In particular, a compact fluorimetric sensor equipped with an LED source and a high-sensitivity PMT detector was implemented for the selective detection of AFM1. The additional use of β-CD as a fluorescence enhancer was investigated for the purpose of improving the instrumental sensitivity. A simple milk clarification procedure was then performed using ethanol followed by a centrifugation step; the clarified sample was analyzed for AFM1 using a fluorimetric assay and the results were compared with those obtained by HPLC analysis, showing good accuracy. The system allowed the detection of 25 ng L^{-1} (LOQ) of AFM1 without sample preconcentration, but when Succinyl-beta-cyclodextrin (β-CD-Su) was used (5 mM), the overall sensitivity of the system was increased to 5 ng L^{-1}.

9.7 Chemosensing: a Molecular Modeling-driven Strategy

It is clear that molecular recognition methods based on chemosensors have several advantages over biological recognition systems, such as low cost, robustness and stability for long storage periods. These points are particularly important in food diagnostics, where a great number of tests at a reasonably low price are required, especially when raw materials such as crops are collected and transferred to transformation industries. On the other hand, the main drawback of chemosensing is the lower selectivity for the guest compared with antibody-based approaches.

In contrast to biomedical applications, food diagnostics is mainly oriented towards the development of rapid but reliable methods to be applied on-site for safety control along the production chain. The main target are residues and xenobiotics regulated by law in raw materials as well as in retail products. Quantitative analysis is usually not required, since the establishment of a reliable cut-off level may allow for the rapid evaluation of bulk material. Positive samples are then further analyzed by routine methods based on laboratory techniques.

Rapid methods based on immunochemical recognition (*e.g.*, lateral flow devices, dipsticks) are widely used for food safety control, despite the limited on-site applicability due to the low stability of antibodies under common environmental conditions. The development of reliable and sensitive rapid methods based on chemosensing to be used on-site or to be automated and integrated in the production chain are considered a gold standard for the agro-food area. Therefore, further studies should be performed to improve this selectivity in order to comply with food safety requirements.

Strong efforts have been made in recent years to replace natural biomolecules with artificial biomimic receptors. Different approaches have been developed, such as combinatorial synthesis of molecular receptors, a combinatorial library of nucleic acids (aptamers) and molecular imprinted polymers (MIPs). These receptors can be designed to have the required specificity for a chosen analyte or a whole class of target analytes.

Since biomimetic techniques allow the problem of the low stability of antibodies, observed in extreme environments such as pH, organic solvents and high temperatures, to be overcome, the development of synthetic receptors based on mimic techniques has also recently been introduced into the field of mycotoxin detection. In particular, many studies have been carried out on the development of MIP-based methods for food diagnostics.[84]

A different approach, based on chemosensing, involves the use of 'smart molecules' showing biomimetic properties towards mycotoxins. These compounds could exhibit a substrate-selective recognition mechanism similar to that of antibodies or enzymes, while being more stable and affordable. The development of such biomimetic compounds requires a strong synthetic effort based on sound molecular design to allow a target-tailored approach.

CDs can be considered a good substrate for the development of suitable chemosensors, on account of their ability to host small compounds and to induce changes in the guest spectral properties such as fluorescence. Moreover, CDs are atoxic, stable under a wide environmental conditions and can be stored for long periods. Hence the implementation of CD-based systems for food diagnostic can potentially lead to the development of sensitive fluorescence-based recognition systems at a reasonably low price. The main drawback to be overcome is the low specificity, which is very important when trace contaminants need to be determined in a very complex matrix such as food.

Assistance in this direction can be gained by application of a computer-aided strategy for chemosensing design. Molecular modeling has been applied to the design of synthetic receptors since it allows computational chemists to generate and refine molecular geometry, driving synthesis with feedback. Molecular modeling is particularly important in two aims: the comprehension of experimental data in the absence of structural information and the design and optimization of new compounds to drive new syntheses. Hence a smart strategy for the improvement of CD selectivity should be based on the scheme shown in Figure 9.5.

The functionalization strategy should be driven by molecular modeling in order to focus the synthetic efforts only on those derivatives which are promising in terms of improved specificity. Then, experimental testing should return important information to improve the computation model and to optimize the docking procedure.

Functionalization strategies may be based on the introduction of chemical groups chosen on the basis of the chemical characteristics of the guest.

An interesting approach may be found in the study of the interaction of food xenobiotics with nuclear receptors. In particular, molecular modeling may help to highlight the interactions occurring inside the binding site between the xenobiotic of interest (the ligand) and the key amino acid residues. This information may be later exploited to design a specific recognition system resembling the natural binding site and thus showing improved specificity towards the target compound.

As an example, recently the Joint FAO–WHO Expert Committee database of xenoestrogens, containing 1500 compounds, was analyzed using an integrated *in silico* and *in vitro* approach.[85] In particular, this analysis identified 31 potential estrogen receptor α ligands that were reduced to 13 upon applying a stringent filter based on ligand volume and binding mode. Among the 13 potential xenoestrogens, four were already known to exhibit estrogenic activity and the other nine were assayed *in vitro*, determining the binding affinity to the receptor and biological effects. The applied approach allowed the characterization of the chemical interactions driving the binding between the ligand and the α-ER binding site. The application of an integrated *in silico* and *in vitro* approach allows us to increase the speed of the analysis of food additives databases for the identification of potential xenoestrogens.

Starting from this evidence, it is possible to design a ligand-tailored chemosensor, by functionalizing a CD with proper groups allowing for the

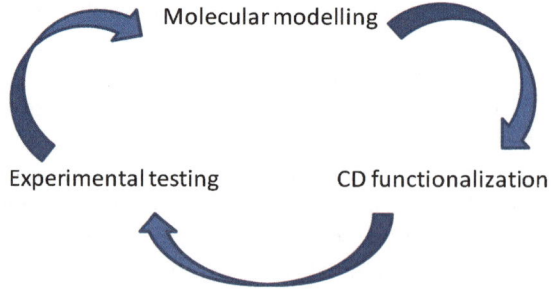

Figure 9.5   Scheme for improvement of CD selectivity.

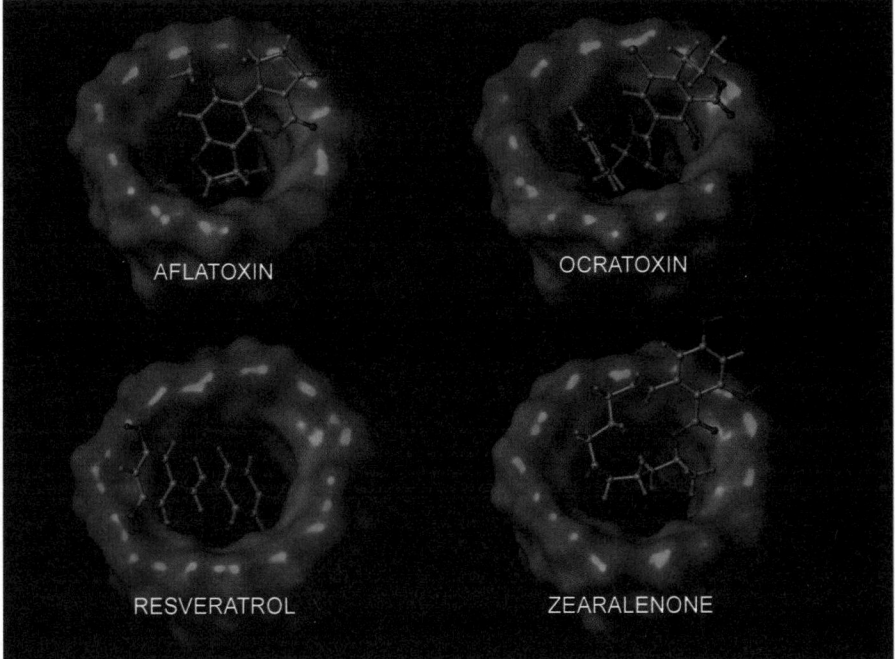

Figure 9.6 An example of inclusion complexes, obtained upon docking analysis, between CDs, the three most important mycotoxins and another biologically relevant molecule.

establishment of polar and/or non-polar bonds. This virtual cycle shows the potential to improve considerably the chemosensing design and development not only for CD-based systems but also for other chemical recognition systems.

In Figure 9.6, an example of inclusion complexes, obtained upon docking analysis, between CDs, three most important mycotoxins and another biologically relevant molecule has been reported.

References

1. S. M. Roberts (ed.), *Molecular Recognition – Chemical and Biochemical Problems*, Royal Society of Chemistry, Cambridge, 1989.
2. J. M. Lehn, *Supramolecular Chemistry*, Wiley-VCH, Weinheim, 1995.
3. E. Fischer, *Ber. Dtsch. Chem. Ges.*, 1984, **27**, 2985.
4. J. M. Lehn, *Angew. Chem. Int. Ed.*, 1990, **29**, 1304.
5. G. W. Gokel (ed.), *Advances in Supramolecular Chemistry*, JAI Press, Greenwich, CT, 1990, Vol. 1.
6. F. Vögtle, *Supramolecular Chemistry – An Introduction*, Wiley, Chichester, 1993.
7. F. Vögtle and E. Weber (eds), *Host–Guest Complex Chemistry – Macrocycles*, Springer, Berlin, 1985.

8. B. Dietrich, P. Viout and J. M. Lehn, *Macrocyclic Chemistry: Aspects of Organic and Inorganic Supramolecular Chemistry*, Wiley-VCH, Weinheim, 1993.
9. D. Hamilton, in *Bioorganic Chemistry Frontiers*, ed. H. Dugas, Springer, Berlin, 1991, Vol. 2, p. 115.
10. J. Szejtly and T. Osa (eds), *Comprehensive Supramolecular Chemistry*, Vol. 3, Cyclodextrins, Elsevier Science, Oxford, 1995.
11. F. Ellouze, N. B. Amar and A. Deratani, *Chimie*, 2011, **14**, 967.
12. W. Saenger, J. Jacob, K. Gessler, T. Steiner, D. Hoffman, H. Sanbe, K. Koizumi, S. M. Smith and T. Tanaka, *Chem. Rev.*, 1998, **98**, 1787.
13. K. L. Larsen, *J. Inclus. Phenom. Macrocycl. Chem.*, 2002, **43**, 1.
14. R. Singh, N. Bharti, J. Madan and S. N. Hiremath, *J. Pharm. Sci. Technol.*, 2010, **2**, 171.
15. J. Szejtli, *Chem. Rev.*, 1998, **98**, 1743.
16. T. Loftsson and M. E. Brewster, *J. Pharm. Sci.*, 1996, **85**,1017.
17. E. Schneiderman and A. M. Stalcup, *J. Chromatogr. B*, 2000, **745**, 83.
18. G. Shmidt, *Trends Biotechnol.*, 1989, **7**, 244.
19. Y. Matsui and K. Mochida, *Bull. Chem. Soc. Jpn.*, 1979, **52**, 2808.
20. H. A. Benesi and L. H. Hildebrand, *J. Am. Chem. Soc.*, 1949, **71**, 2703.
21. M. Suzuki and Y. Sasaki, *Chem. Pharm. Bull.*, 1984, **32**, 832.
22. I. M. Brereto, T. M. Spotswood, S. Lincoln and E. H. Williams, *J. Chem. Soc., Faraday Trans. 1*, 1984, **80**, 3147.
23. P. E. Hansen, H. D. Dettman and B. D. Sykes, *J. Magn. Reson.*, 1985, **62**, 487.
24. S. E. Brown, J. H Coates, S. F. Lincoln, D. R. Coghlan and C. J. Easton, *J. Chem. Soc., Faraday Trans.*, 1991, **87**, 2699.
25. A. F. Casy, *Trends Anal. Chem.*, 1993, **12**, 185.
26. V. I. Sokov, V. L. Bondareva and G. V. Shustov, *Metallorg. Khim.*, 1991, **4**, 697 (in Russian).
27. K. Kano, K. Yoshiyasu, H. Yasuoka, S. Hata and S. Hashimoto, *J. Chem. Soc., Perkin Trans. 2*, 1992, 1265.
28. A. Ueno, Q. Chen, I. Suzuki and T. Osa, *Anal. Chem.*, 1992, **64**, 1650.
29. H. Schönherr, M. W. J. Beulen, J. Huskens, F. C. J. M. van Veggel, D. N. Reynhoudt and G. J. Vancso, *J. Am. Chem. Soc.*, 2000, **122**, 4963.
30. S. Zapotoczny, T. Auletta, M. R. de Jong, H. Schönherr, J. Huskens, F. C. J. M. van Veggel, D. N. Reynhoudt and G. J. Vancso, *Langmuir*, 2002, **18**, 6988.
31. T. Auletta, M. R. de Jong, A. Mulder, J. Huskens, F. C. J. M. van Veggel, D. N. Reynhoudt, S. Zou, S. Zapotoczny, H. Schönherr, G. J. Vancso and L. Kuipers, *J. Am. Chem. Soc.*, 2004, **126**, 1577.
32. M. A. Martin, A. I. Olives, B. del Castillo and J. C. Menendez, *Curr. Pharm. Anal.*, 2008, **4**, 106.
33. F. Cramer, *Chem. Ber.*, 1951, **81**, 851.
34. K. Ikeda, K. Vekama and M. Otaqiri, *Chem. Pharm. Bull.*, 1975, **23**, 201.
35. M. Vikmon, A. Stadler-Szoke and J. Szejtli, *J. Antibiot.*, 1985, **38**, 1822.
36. S. O. Fakayode, M. Lowry, K. A. Fletcher, X. Huang, A. M. Powe and I. M. Warner, *Curr. Anal. Chem.*, 2007, **3**, 171.
37. J. Szejtli, *Cyclodextrin Technology*, Kluwer, Dordrecht, 1988.

38. S. Allenmark, *Chirality*, 2003, **15**, 409.
39. B. D. Wagner, *Curr. Anal. Chem.*, 2007, **3**, 183.
40. T. Kinoshita, F. Iinuma and A. Tsuji, *Biochem. Biophys. Res. Commun.*, 1973, **51**, 666.
41. K. Kano, I. Takenoshita and T. Ogawa, *J. Phys. Chem.*, 1982, **86**, 1833.
42. N. J. Turro, G. S. Cox and X. Ki, *Photochem. Photobiol.*, 1983, **36**, 149.
43. H. Kondo, H. Nakatani and K. J. Hiromi, *Biochemistry*, 1976, **79**, 393.
44. K. W. Street, *J. Liq. Chromatogr.*, 1987, **10**, 655.
45. S. Hashimoto and J. K. Thomas, *J. Am. Chem. Soc.*, 1985, **107**, 4655.
46. L. J. Cline-Love, M. L. Grayeski, J. Noroski and R. Weinberger, *Anal. Chim. Acta*, 1985, **170**, 3.
47. R. Temia, S. Scypinski and L. J. Cline-Love, *Environ. Sci. Technol.*, 1985, **19**, 155.
48. J. P. Alarie and T. Vo Dinh, *Talanta*, 1991, **38**, 529.
49. E. M. Martin Del Valle, *Process Biochem.*, 2004, **39**, 1033.
50. G. Astray, C. Gonzalez-Barreiro, J. C. Mejuto, R. Rial-Otero and J. Simal-Gandara, *Food Hydrocolloids*, 2009, **23**, 1631.
51. G. Cravotto, A. Binello, E. Baranelli, P. Carraro and F. Trotta, *Curr. Nutr. Food Sci.*, 2006, **2**, 343.
52. T. Loftsson, and M. E. Brewster, *J. Pharm. Pharmacol.*, 2010, **62**, 1607.
53. J. Szejtli, *Cyclodextrins and Their Inclusion Complexes*, Akadémiae Kiadó, Budapest, 1982.
54. Y. Okabe, H. Yamamura, K. I. Obe, K. Otha, M. Kawai and K. Fujita, *J. Chem. Soc., Chem. Commun.*, 1995, 581.
55. A. Hybl, R. E. Rundle and D. E. William, *J. Am. Chem. Soc.*, 1965, **87**, 2779.
56. W. Saenger, M. Noltemeyer, P. C. Manor, B. Himgerty and B. Klar, *Bioorg. Chem.*, 1976, **5**, 187.
57. D. Rong and V. T. DSouza, *Tetrahedron Lett.*, 1990, **31**, 4275.
58. A. P. Croft and R. A. Bartsch, *Tetrahedron*, 1983, **39**, 1417.
59. G. Guebitz and M.G. Schmid, *Chiral Separations by Capillary Electrophoresis, Chromatography Science Series*, Edited by A. Van Eeckhaut, Y. Michotte. Vol. 100, CRC Press, Boca Raton, FL, 2010, pp. 47–85.
60. F. Kitagawa and K. Otsuka, *J. Chromatogr. B*, 2011, **879**, 3078.
61. M. L. Vazquez, A. Cepeda, P. Prognon, G. Mahuzier and J. Blaus, *Anal. Chim. Acta*, 1991, **255**, 343.
62. A. Cepeda, C. M. Franco, C. A. Fente, B. I. Vázquez, J. L. Rodríguez, P. Prognon and G. Mahuzier, *J. Chromatogr. A*, 1996, **721**, 69.
63. C. Dall'Asta, G. Ingletto, R. Corradini, G. Galaverna and R. Marchelli, *J. Inclus. Phenom. Macrocycl. Chem.*, 2003, **45**, 257.
64. A. Amadasi, C. Dall'Asta, G. Ingletto, R. Pela, R. Marchelli and P. Cozzini, *Bioorg. Med. Chem.*, 2007, **15**, 4585.
65. M. Aghamohammadi and N. Alizadeh, *J. Lumin.*, 2007, **127**, 575.
66. G. Ramírez-Galicia, R. Garduño-Juárez and M. Gabriela Vargas, *Photochem. Photobiol. Sci.*, 2007, **6**, 110.

67. C. Dal,lAsta, A. Faccini, G. Galaverna, R. Corradini, A. Dossena and R. Marchelli, *J. Inclus. Phenom. Macrocycl. Chem.*, 2009, **64**, 331.
68. R. Verrone, L. Catucci, P. Cosma, P. Fini, A. Agostiano, V. Lippolis and M. Pascale, *J. Inclus. Phenom. Macrocycl. Chem.*, 2007, **57**, 475.
69. O. J. Francis Jr, G. P. Kirschenheuter, G. M. Ware, A. S. Carman and S. S. Kuan, *J. Assoc. Off. Anal. Chem.*, 1988, **71**, 725.
70. C. M. Franco, C. A. Fente, B. I. Vázquez, A. Cepeda, G. Mahuzier and P. Prognon, *J. Chromatogr. A*, 1998, **815**, 21.
71. M. L. Vazquez, C. A. Fente, C. M. Franco, A. Cepeda, G. Mahuzier and P. Prognon, *Anal. Commun.*, 1999, **36**, 5.
72. R. D. Holland and M. J. Sepaniak, *Anal. Chem.*, 1993, **65**, 1140.
73. J. Wei, E. Okerberg, J. Dunlap, C. Ly and J. B. Shear, *Anal. Chem.*, 2000, **72**, 1360.
74. P. Golinski and J. Chelkowski, *J. Assoc. Off. Anal. Chem.*, 1978, **61**, 586.
75. D. C. Hunt, L. A. Philip and N. T. Crosby, *Analyst*, 1979, **104**, 1171.
76. C. M. Maragos and M. Appell, *J. Chromatogr. A*, 2007, **1143**, 252.
77. C. M. Maragos, M. Appell, V. Lippolis, A. Visconti, L. Catucci and M. Pascale, *Food Addit. Contam. A*, 2008, **25**, 164.
78. C. A. Fente, J. J. Ordaz, B. I. Vázquez, C. M. Franco and A. Cepeda, *Appl. Environ. Microbiol.*, 2001, **67**, 4858.
79. H. K. Abbas, R. M. Zablotowicz, M. A. Weaver, B. W. Horn, W. Xie and W. T. Shier, *Can. J. Microbiol.*, 2004, **50**, 193.
80. T. Rojas-Durán, I. Sánchez-Barragán, J. M. Costa-Fernández and A. Sanz-Medel, *Analyst*, 2007, **132**, 307.
81. T. R. Rojas-Durán, C. A. Fente, B. I. Vázquez, C. M. Franco, A. Sanz-Medel and A. Cepeda, *Int. J. Food Microbiol.*, 2007, **115**, 149.
82. C. Cucci, A. G. Mignani, C. Dall'Asta, R. Pela and A. Dossena, *Sens. Actuators B*, 2007, **126**, 467.
83. A. G. Mignani, L. Ciaccheri, C. Cucci, A. A. Mencaglia, A. Cimato, C. Attilio, H. Ottevaere, H. Thienpont, R. Paolesse, M. Mastroianni, D. Monti, M. Gerevini, G. Buonocore, M. A. Del Nobile, A. Mentana, M. F. Grimaldi, C. Dall'Asta, A. Faccini, G. Galaverna and A. Dossena, *IEEE Sens. J.*, 2008, **8**, 1342.
84. K. Mosbach and O. Ramstrom, *Bio/Technology*, 1996, **14**, 163.
85. A. Amadasi, A. Mozzarelli, C. Meda, A. Maggi and P. Cozzini, *Chem. Res. Toxicol.*, 2009, **22**, 52.

Subject Index

References to figures are given in *italic* type. References to tables are given in **bold** type.